D1547164

ELEMENTARY METALLURGY

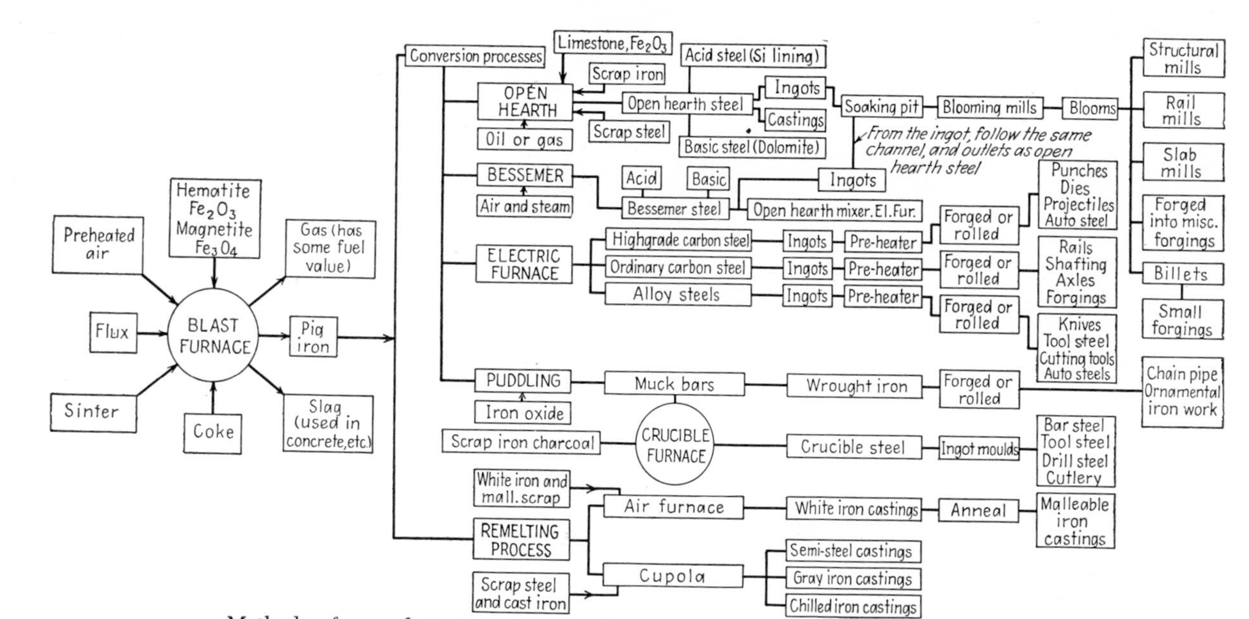

Methods of manufacturing commercial iron and steel. (*Drawn by W. M. Johnson.*)

ELEMENTARY METALLURGY

by W. T. FRIER, B.S.

Part-time Instructor in Metallurgy
Pennsylvania State College Extension
Formerly at General Electric Technical
Night School, Erie Works

SECOND EDITION

McGRAW-HILL BOOK COMPANY, INC.

New York Toronto London

1952

ELEMENTARY METALLURGY

Library of Congress Catalog Number: 51-12607

XI

22419

THE MAPLE PRESS COMPANY, YORK, PA.

PREFACE TO THE FIRST EDITION

This text is a compilation of notes that the author has used for several years in teaching the subject of metallurgy in the General Electric Technical Night School of the Erie Works, General Electric Company, Erie, Pa.

The aim is to present the subject in such a manner as to be easily understood by students in technical institutes, war production courses, vocational high schools, and the like, even though they may have little preliminary knowledge of metallurgy or chemistry. If this book helps readers better to understand the characteristics of the metals they are handling, or especially if it awakens an interest in metals that will lead to further study, the author will feel well repaid for his efforts.

Considerable space has been devoted to constitution diagrams because the author feels that these are fundamental and should be firmly entrenched in the mind. It must be remembered that these diagrams are not theoretical; they are the result of observed behavior according to natural laws and have been built up by a vast amount of experimental work involving cooling curves, electrical conductivity, dilation, etc. The author believes, too, that all the theories of hardening—by alloying, quenching, precipitation hardening, cold-working, and even the strength of grain boundaries—can be presented as cases of strained atomic spacing, deviations from the natural spacing.

All temperatures have been given in the Fahrenheit scale, as that is the one most used in practical work.

It is earnestly recommended that students be given an opportunity to see motion pictures of blast furnaces, steelmaking processes, and rolling mills during the course. Such films are easily obtainable from the U.S. Bureau of Mines and from the various steel companies. Also, if possible, students should see physical tests being made on metals and should be instructed in the use of the microscope.

The author wishes to thank the following individuals for help in the preparation of the text:

Mr. J. R. Brown, Former Supervisor of Education, Erie Works, General Electric Company.

Mr. Otto Hiller of the works laboratory, General Electric Company.

Mr. John Clarke, Assistant Superintendent of Foundry Division, Erie Works, General Electric Company, for help with the chapter on Nonferrous Alloys.

Dr. E. C. Bain, United States Steel Corporation, for assistance and encouragement, especially in connection with austempering and the S curve.

W. T. FRIER

ERIE, PA.
August, 1942

PREFACE TO THE SECOND EDITION

Since the publication of the first edition of this book nine years ago, such extensive progress in the field of metallurgy has been made that a complete revision of the text has become necessary.

This revision has given the author an opportunity to introduce new material as well as the opportunity to rearrange some of the old material in a more logical order, for example, placing "forming of metals" after information on their properties and including all the steelmaking processes in one chapter. A list of helpful motion pictures has been added, and some obscure points have been clarified as a result of the criticisms made by teachers who used the first edition.

The purpose of the book remains the same, but the author regrets that he could not treat some subjects more extensively. A fuller treatment would have made the book too large for the purpose intended. A list of pertinent references has been placed at the end of each chapter for students who wish to do further reading in the field of metallurgy.

W. T. FRIER

ERIE, PA.
January, 1952

CONTENTS

Chapter 1
IRON

Source

Iron and all the metals we shall study in this book are found in nature as either oxides or sulfides, which are known as *ores.* An ore may be defined as a mineral from which one or more elements may be extracted with profit. Practically all our iron is extracted from iron oxide ore. If a piece of ordinary iron or steel is left out in the weather, it rusts, which means that it is returning to the state in which it was found, for rust is simply iron oxide. Red iron ore is this same oxide with varying percentages of earth, sand, or rock, due to its having been taken from the ground.

Chemical Analysis

The chemical formula for this red iron ore is Fe_2O_3. By the law of combining weights, this means that the ore contains 2×55.84 parts by weight of iron to 3×16 parts by weight of oxygen. (All elements combine in certain definite proportions according to their atomic weights, that of iron being 55.84 and of oxygen 16.) The percentage of iron in Fe_2O_3, therefore, is calculated to be 70 per cent. (Another ore of iron, Fe_3O_4, is calculated to be nearly 72 per cent.) Because of the earthy material that would naturally be associated with the iron oxide as it is mined, however, the percentage of iron is usually between 50 and 55 per cent. This earthy material is known as *gangue.*

PRODUCTION PROCESSES

Winning

The first and principal step in winning iron (or any other metal) from its ore is to remove the oxygen that is in combination with it. Such a process is known as *reduction*. While the chem-

Fig. 1. Electrical equipment used at the Wabigon open-pit mine on the Mesabi Range in northern Minnesota, Hanna Ore Mining Company. (*Courtesy General Electric Company.*)

ist has a more elaborate definition,[1] it is sufficient in this course to consider *reduction* as a reduction of the oxygen content of a compound and to consider *oxidation* (the opposite of reduction) as an increase in the oxygen content.

[1] Reduction is the increase in valence of an element or radical in the negative direction (as a result of gaining or borrowing electrons), while oxidation, conversely, denotes an increase in valence in the positive direction (as a result of losing or sharing electrons).

FIG. 2. Close-up view of blast furnace, showing skip hoist for charging materials, large "down-comer" pipes for taking off gases, etc. (*Courtesy The Jackson Iron and Steel Company, Jackson, Ohio, through Bradley Booth, metallurgist.*)

The Blast Furnace

Practically all iron is produced in a blast furnace (see cross section, Fig. 3). This is essentially an apparatus for heating the ore in the presence of an excess of carbon, principally in the form of carbon monoxide, which is the *reducing agent.* That is, it combines with the oxygen present in the ore and thus

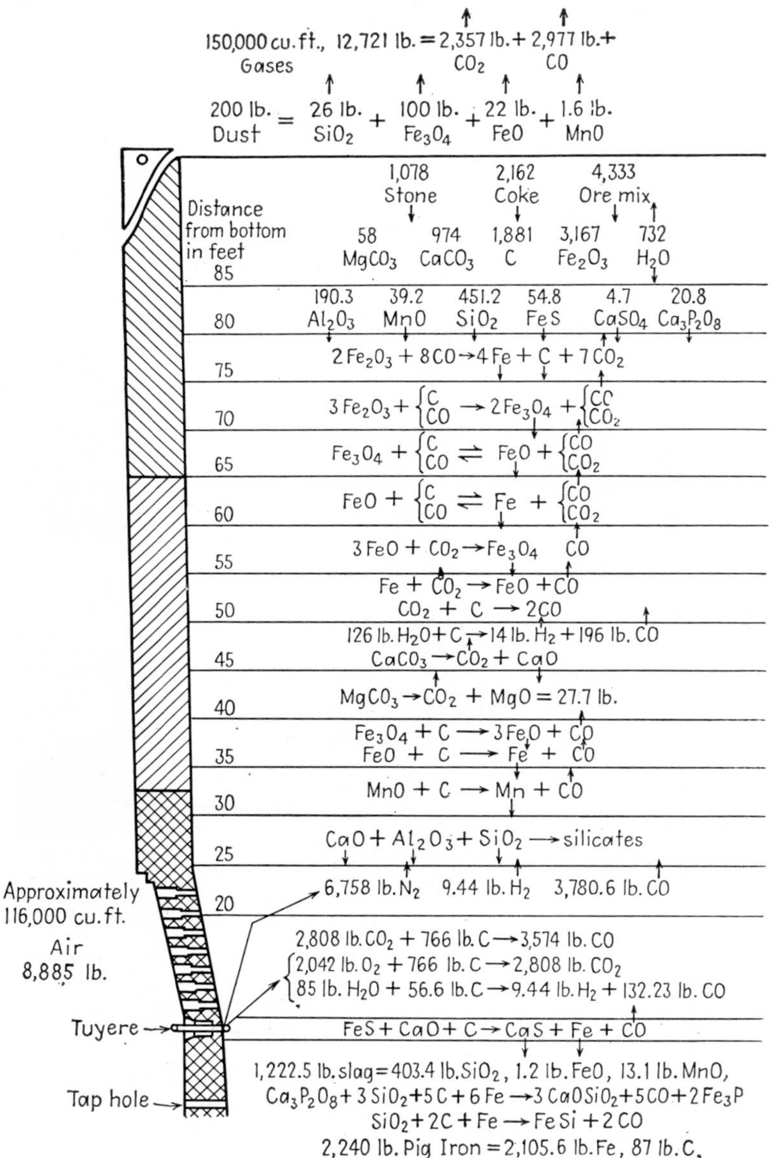

FIG. 3. The making of a ton of pig iron. A diagram showing the raw materials and the products of the blast furnace, their relative weights, and the changes that take place therein. (*Reprinted by permission from J. M.*

amount of limestone must be carefully calculated. The ratio of total acids[1] to bases in the slag is slightly over 1 to 1, for the reason that the furnace is lined with silica brick and an excess of lime would attack the lining. The lime also removes some sulfur, which goes into the slag as CaS. As has been stated, the production of iron in the blast furnace is a reduction process. It is therefore only natural to suppose that other elements besides iron would also be reduced, and that is true. Note these equations for reactions in the lower part of the furnace:

$$Ca_3P_2O_8 + 3SiO_2 + 5C + 6Fe \rightarrow 3CaOSiO_2 + 5CO + 2Fe_3P$$
$$SiO_2 + 2C + Fe \rightarrow FeSi + 2CO$$
$$MnO + C \rightarrow Mn + CO$$

Thus the elements so set free are present in the pig iron, along with sulfur and carbon, to the extent of being impurities; and they must be removed to a great extent in subsequent refining processes such as the making of steel.

Besides the pig iron, another product leaving the furnace is *slag*, whose composition is shown in the fourth line from the bottom of Fig. 3. While of no further use to the metallurgist, slag has a multitude of uses, such as road-building material, railroad ballast, fertilizer, and in cement manufacture.

Note that the gases leaving the furnace contain more than 20 per cent CO. This CO has considerable heating value; the gas is cleaned and finds many uses around the plant, one of the principal uses being the preheating of the air that is blown into the furnace. This is accomplished in "stoves" (see Fig. 4), which are brick-lined steel shells built up inside with brick checkerwork and arranged so that they are first heated by the burning gas and then connected into the air supply, which is

[1] Alumina (clay, for instance) must be taken into account in calculating a furnace charge, although it is neither a strong base nor acid. Magnesia, if present, is calculated as a strong base.

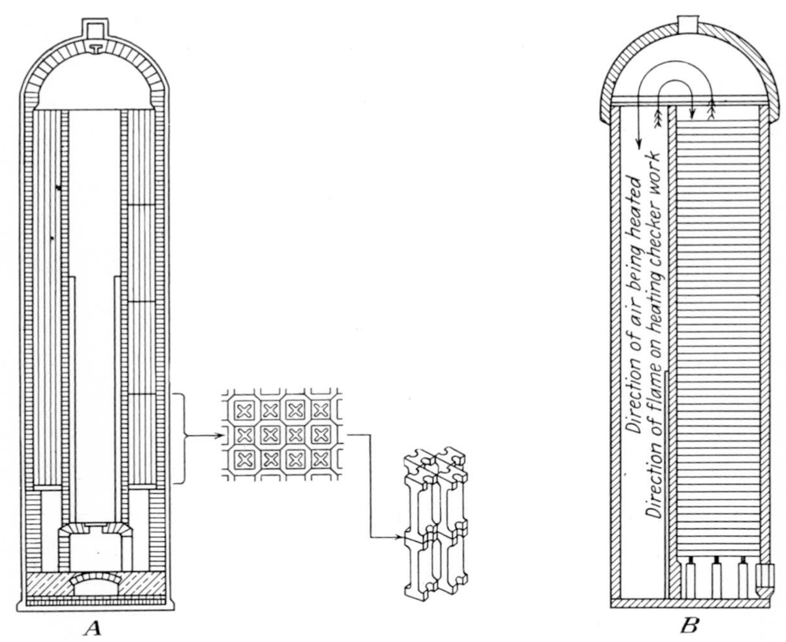

FIG. 4. Blast-furnace stoves, linings, and checkers. *A*, longitudinal section—two-pass, center-combustion type. *B*, longitudinal section—two-pass, side-combustion type. (*Reprinted by permission from the J. M. Camp and C. B. Francis, "The Making, Shaping and Treating of Steel, 5th ed., Carnegie-Illinois Steel Corporation.*)

thus heated to about 1400°F. Other ways of utilizing the heating value of blast-furnace gas include mixing it with about three parts coke-oven gas to heat coke ovens and soaking pits and using it unmixed in the power plant. A few plants use it to run gas engines for blowing the air supplied to the furnace.

DEFINITIONS

An **ore** may be defined as a mineral from which one or more elements can be extracted with profit.

Reduction may be defined as a reduction of the oxygen content of a compound.

Oxidation may be defined as an increase in the oxygen content of a compound.

QUESTIONS

1. What materials are charged into the blast furnace?
2. What materials are taken out of the blast furnace?
3. Of the 2 tons of ore charged into the furnace, 1 ton leaves the furnace as iron. How does the other ton leave?
4. As the charged ore descends, what reducing agent acts on it first?
5. Name some impurities that are reduced along with the iron.

REFERENCES

CAMP, J. M., and C. B. FRANCIS, "The Making, Shaping and Treating of Steel," 5th ed., Carnegie-Illinois Steel Corporation, Pittsburgh, 1940.
"Metals Handbook," pp. 315–319, American Society for Metals, Cleveland, 1948.
STOUGHTON, BRADLEY, "The Metallurgy of Iron and Steel," 4th ed., McGraw-Hill, New York, 1934.
SWEETSER, R. H., "Blast Furnace Practice," McGraw-Hill, New York, 1938.
TEICHERT, E. J., "Ferrous Metallurgy," 2d ed., vol. I, McGraw-Hill, New York, 1944.

Chapter 2
STEEL

It was brought out in the previous chapter that the product of the blast furnace, pig iron, is so impure that it finds no use in that condition. The next two chapters will describe refining processes used to convert the pig iron into the finished steel, cast iron, and wrought iron.

THE BESSEMER CONVERTER

Since the impurities entered the iron as a result of reduction, it is only natural to infer that they can be removed by oxidation. To a great extent this is true. The remaining processes that refine the pig iron to cast iron and steel are based on oxidation. The simplest of these processes is the *bessemer converter*, which is the simplest in chemistry and the simplest in operation. It is also the most economical process, for it uses air as the oxidizing agent; and for fuel to furnish the necessary heat, it makes use of the very impurities that are to be *burned out.*

Bessemer Reactions

Notice how, in the sketch (see Fig. 5), the converter, tilted on its side, is receiving a charge of molten pig iron. The next step is to tip it back to vertical and to turn on a blast of air, which enters through the $\frac{3}{8}$-in. holes at the bottom. The air is blown in cold, and yet the temperature immediately rises 200°F. and keeps rising for several minutes, the fuel within the iron burning as the air reaches it.

In fact, the oxygen attacks the iron first. This is because there is such a large proportion of iron present. However, as the FeO so formed diffuses through the metal, it in turn oxidizes the silicon and manganese, because these two elements will combine with FeO, forming SiO_2 and MnO. All three of these reactions

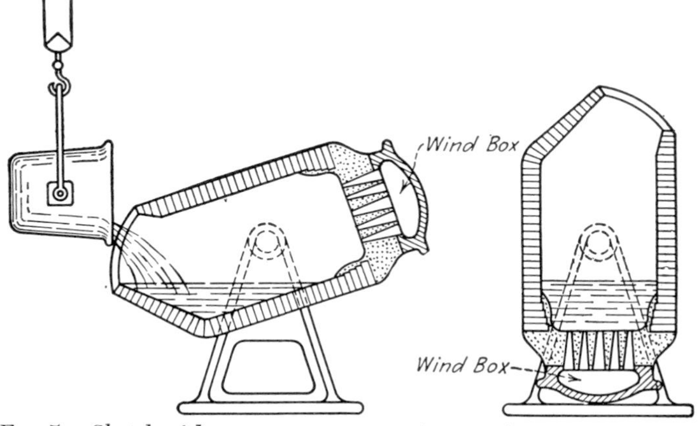

Fig. 5. *Sketch of bessemer converter.* (*Drawn by P. R. Schilling.*)

produce heat and rapidly increase the temperature of the molten metal to the point where the carbon begins to burn. As the carbon burns, it furnishes one of the most spectacular sights of the steel industry—a long yellow flame 30 ft. high. Unfortunately, this flame is more spectacular than useful, for the carbon is burned only to CO within the bath. The final combustion, CO to CO_2, takes place outside the converter; and thus only one-third of the heat in the carbon is actually used in heating the metal.

The *blow* must be stopped just before the carbon is completely burned, a matter of only 11 or 12 min.—or the iron itself will be oxidized. As the exact time of stopping the blow is a matter of seconds, very experienced operators are required. At the end of the blow, some manganese is added, as ferromanganese, which is about 80 per cent Mn and 20 per cent Fe. Because manganese has a greater affinity for oxygen than iron has, the following reaction takes place: $FeO + Mn = MnO + Fe$. So important is this addition of Mn that the bessemer process is absolutely dependent on it for the removal of iron oxide, the presence of which would weaken the steel. Not only is the Mn a *deoxidizer*, it is also a *desulfurizer*, as represented by $Mn + FeS = MnS + Fe$.

Some carbon is also added in the form of coke, hard coal, or graphite, the amount depending upon the desired carbon content of the finished steel.

Although the actual blowing takes but 12 min., there are also the emptying and filling to consider; therefore, with the usual furnace capacity of 18 tons, about 50 tons are produced per hour in one converter. Even that is fast; and so, considering also the fuel economy, the question, why make steel any other way? naturally arises.

The fact is that only about 5 per cent of the steel produced in the United States is made by this process. The very speed by which it is made is one great disadvantage, because the control that is possible with the slower processes is lacking. Besides, on account of the 79 per cent of nitrogen in the air that is blown in, the steel often absorbs enough of this gas to injure its quality.[1] Bessemer steel therefore is usually considered to be of not quite so good quality as open-hearth steel, which makes up about 90 per cent of our steel.

[1] Not only has nitrogen a tendency to harden steel; but steel, when liquid, also absorbs much more gas than it can retain after it solidifies. The residual gas is thrown out as bubbles, many of which may remain as holes.

Two more disadvantages of the bessemer process are that the converter cannot handle steel scrap except to a very limited extent, nor can this process as used in the United States remove sulfur or phosphorus. As these elements are injurious to steel in most cases,[1] ores containing them cannot be used, and the supply of other ores is constantly decreasing.

Duplexing

There is one process, a combination process, which takes advantage of the rapid purification of the bessemer process and eliminates the other disadvantages. It is known as *duplexing* and consists in oxidizing most of the carbon and all of the silicon and manganese in the converter and then transferring the product to the slower basic open-hearth furnace, where the phosphorus and remainder of the carbon are removed. The basic open-hearth furnace will be described presently. No basic bessemer is operated in the United States. All converter linings are siliceous (acid). This is because we still have a supply of ores low enough in phosphorus that good steel can be produced without its removal, while European ores are so high in phosphorus that it becomes a by-product, used as fertilizer.

Chemical Factors in Refining

The statement in the previous chapter that sulfur and phosphorus (objectionable elements in practically all steels) cannot be removed in the bessemer process as carried out in this country calls for additional discussion in regard to acid-base reactions before describing the basic open-hearth process of steelmaking.

Anyone who has studied chemistry is familiar with such equations as $H_2SO_4 + Ca(OH)_2 = 2H_2O + CaSO_4$, meaning that

[1] One exception is in the case of screw stock; for instance, wood screws do not require a great deal of strength. Also, a small amount of sulfur in the metal aids machining.

when sulfuric acid reacts with a solution of lime, neither one survives, but calcium sulfate and water form.

The reactions by which sulfur and phosphorus are removed are similar. Of course, there could be no water present at the high temperature of molten iron or steel. At such temperatures practically all the elements present are in their oxide state. But they are *liquid,* and so the reactions can proceed as in water solution. We speak of:

Acids (OXIDES OF NONMETALS)	Bases (OXIDES OF METALS)
SiO_2	FeO
SO_2	MnO
P_2O_5	CaO
	MgO
	CrO

and we realize the fact that molten lime floating on top of a bath of iron can easily join with and remove the acids SO_2 and P_2O_5.

There is another important fact to consider in the open-hearth process. The most common and by far the cheapest furnace lining is ordinary firebrick; but this is largely SiO_2, which happens to be one of the most active acids in this high-temperature chemistry. It can be easily seen that molten lime (basic) would attack the acid SiO_2 in the lining and "eat," or dissolve, it very quickly.

Any furnace, therefore, in which sulphur and phosphorus removal is attempted must have a *basic* lining. It must be the *oxide* of some *metal.* The one most commonly used is magnesite, which is MgO. Dolomite, which is a combination of magnesium and calcium carbonates (becoming CaO and MgO under heat), is often used; and a very good, but quite expensive, lining is chromite, a combination of iron and chromium oxides.

With this explanation of what is meant by *acid* and *basic* steels, it is clear that the ore for the latter does not need to be free from phosphorus and sulfur. *Acid* steel, however, is likely to be high in these elements unless the ore is carefully selected. In this country all bessemer converters make *acid* steel.[1] The operation is over so quickly that there is not time for much treatment inside the converter.

THE OPEN-HEARTH FURNACE

Most of the steel made in the open-hearth furnaces, however, is basic. Open-hearth furnaces vary in capacity from 50 to 400 tons. A 100-ton furnace is about 40 ft. long and 15 ft. wide. Unlike the bessemer, it requires outside fuel; formerly most open-hearth furnaces used gas for fuel, but this is being extensively replaced by oil. In order to obtain a high enough temperature, the air, and often the gas, is preheated. This is accomplished in brick checkerwork chambers built under the charging floor, the checkerwork resembling that used in the air-preheating "stoves" connected with the blast furnace.

The mode of operation using gas (four chambers) may be seen by referring to the sketch (see Fig. 8). Suppose the air was entering through port *A* and the gas through *B*. (Gas always enters through the lower port, since gas, being lighter, rises to mix with the air.) As the hot waste gases leave the furnace through *C* and *D*, they heat up the brick-checkerwork chambers *E* and *F* on their way to the stack through the reversing valves. This is continued for about 15 to 20 min., and then the valves are reversed so that the air and gas enter the furnace through *C* and *D* and are preheated to 2200°F. in chambers *E* and *F*. Chambers *K* and *M* are now being heated by the hot waste gases for the next cycle.

[1] In the *basic bessemer* process as used in Germany, the metal is poured into a basic-lined ladle and there treated with calcium carbonate or sodium carbonate, usually the latter.

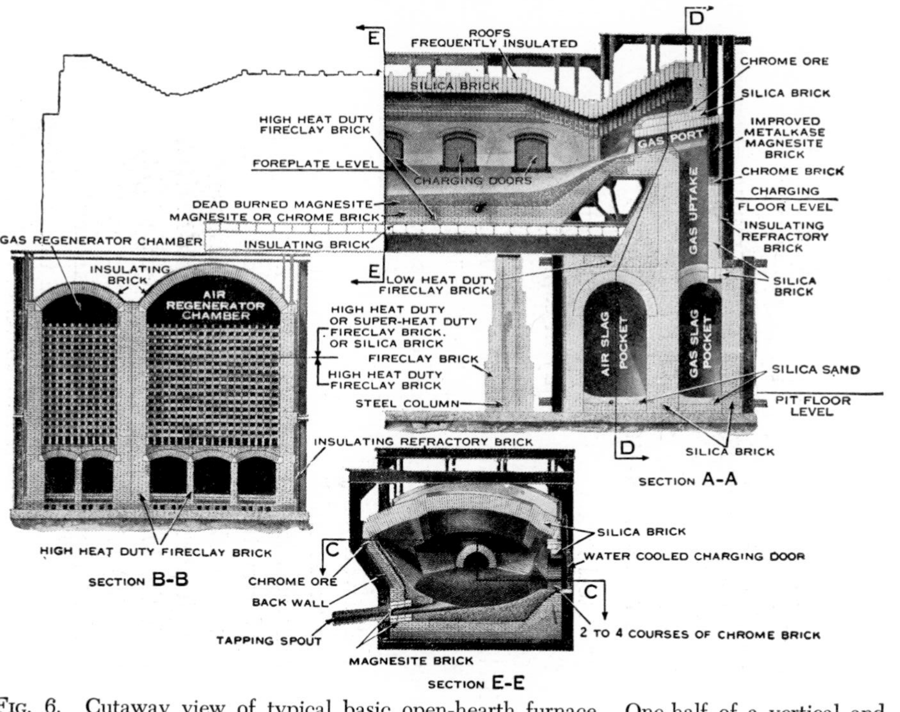

FIG. 6. Cutaway view of typical basic open-hearth furnace. One-half of a vertical and lengthwise section is shown in *AA* (dotted line outlines the other half). Notice that basic bricks and lining are used wherever slag contacts, while roof is made of silica brick. (*Courtesy Harbison-Walker Refractories Company.*)

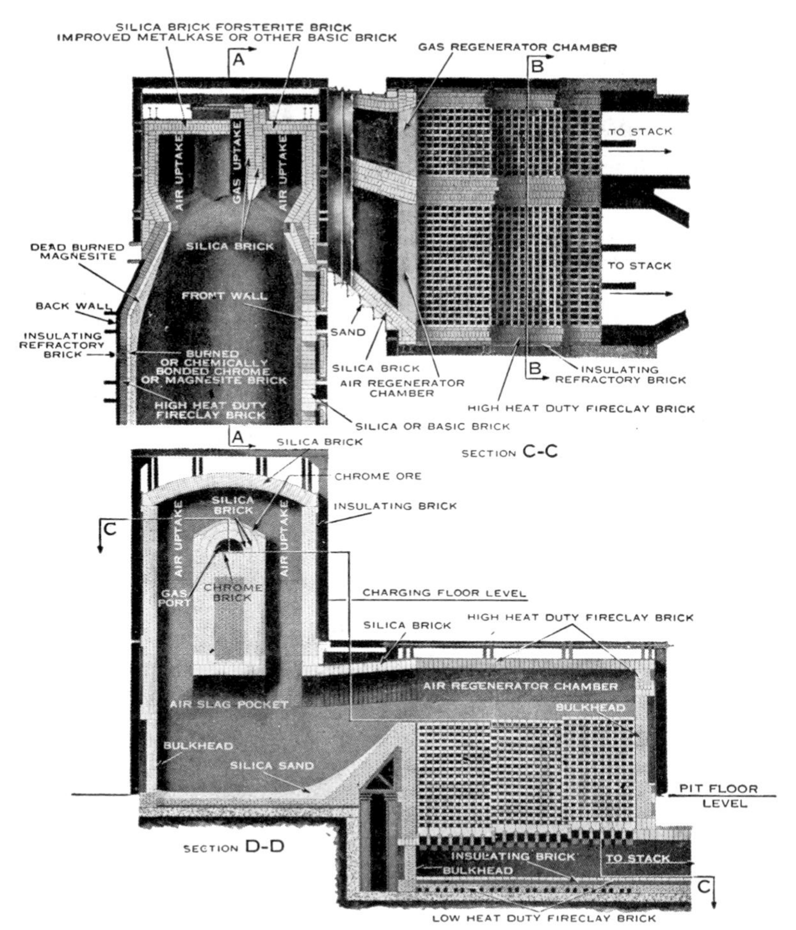

FIG. 7. Continuation of Fig. 6. (*Courtesy Harbison-Walker Refractories Company.*)

Of course, with oil used as the fuel, only two preheating chambers are necessary, and they preheat the air. A great deal of heat is saved by this *regenerative* principle, as it is called. Another method of conserving heat from going to waste up the stack is to insert *waste-heat boilers* whenever it is possible to use

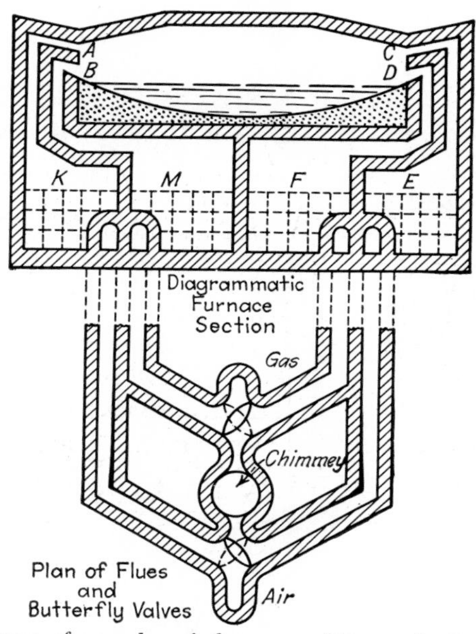

FIG. 8. Diagram of open-hearth furnace. (*Drawn by P. R. Schilling.*)

low-pressure steam to advantage. Draft fans are required, placed between the boiler and the stack. However, in regard to preheating air for the open-hearth furnace, fuel economy is not its primary purpose. The high temperature could not be attained in any other way. In fact, the practical limit to the temperature that can be reached is the degree that the bricks can stand. It is the bricks of the roof that melt first, because heat rises. It has been stated that the charge receives more heat by reflection from the roof than from actual contact with the flame.

The roof can be constructed of silica brick, but the walls and bottom must be made of some basic refractory because they are in contact with the lime in the slag.

In charging the open-hearth furnace, the calculated amount of limestone is placed at the bottom. However, many plants find

FIG. 9. Charging machine placing charges into open-hearth furnaces. (*Courtesy Bethlehem Steel Company.*)

it advantageous first to place a light layer of scrap on the bottom in order to keep the limestone from sticking when the charge is melting. Above the limestone the scrap is placed. From 35 to 50 per cent may be taken as an average amount, although the proportion may go as high as 75 per cent.[1] The variation depends upon the price of scrap, availability of hot metal, and

[1] The enormous amount of steel scrap available can be partly realized when it is known that only 70 to 90 per cent of the steel ingot goes into the final product, and only 50 per cent in the case of castings. Added to this is the great amount of "junk," such as worn-out railroad rails.

whether flush or scrap heats are being made. The pig iron is poured in after the scrap has become hot enough to begin to melt; or if cold pig is used, it is charged with the scrap.

About 9 hr. is required for melting down a 100-ton heat. (The word *heat* as used in connection with melting furnaces means any individual charge as melted.)

Because the melting proceeds so slowly, there is ample time for accurate control by chemical analysis; and usually, the carbon elimination is stopped before that element is all burned out, as in the case of bessemer converter.

Reactions

The oxygen of the blown air was the oxidizing agent in the bessemer process. In the open-hearth process, the oxygen is added principally in solid form, *i.e.*, as *iron oxide*—being either iron ore or roll scale. Its quantity is calculated according to the chemist's analysis in regard to how much silicon, carbon, or other elements are to be removed. Some ore may be added initially along with the limestone, and more is added as needed after the charge has melted. It might be mentioned that appreciable amounts of iron oxide are formed during the melting period from the action of the furnace gases on the metal. During this period the gases impinge directly upon the metal; there is no slag to protect the metal, and good combustion requires some excess oxygen. The reactions at this time can be expressed as

$$2Fe + O_2 = 2FeO$$

$$Fe + CO_2 = FeO + CO$$

and

$$Fe + H_2O = FeO + H_2$$

The next set of reactions, those concerned with the oxidation of impurities by both ore and the FeO just mentioned, are similar to the corresponding reactions in the bessemer converter and are here given in more detail:

$$FeSi + 2FeO \rightarrow SiO_2 + 3Fe$$
$$Mn_3C + 4FeO \rightarrow 3MnO + 4Fe + CO$$
$$2Fe_3P + 5FeO \leftrightharpoons P_2O_5 + 11Fe$$

and

$$Fe_3C + FeO \rightarrow CO + 4Fe$$

These reactions produce heat just as they do in the bessemer process. The gases given off in this period give it the name *ore boil.* The next period is known as the *lime boil* and is caused by the breaking up of the limestone into lime and carbon dioxide gas, the latter bubbling up through the melt and the slag, for by this time the bath of metal is completely under cover of slag, and largely dependent on reactions between slag and metal for control. It is the lime in the slag which removes the phosphorus:

$$2P + 5FeO + 3CaO = 3CaO \cdot P_2O_5 + 5Fe$$

Some sulfur is also removed, according to $FeS + CaO \rightleftharpoons CaS + FeO$; but since this reaction is reversible, complete removal is possible only in the electric furnace, as will be explained under that head. The final period in open-hearth operation is known as the *working period,* which consists in the removal of phosphorus, adjustment of carbon content, and raising the temperature of the bath. Some final adjustment, such as deoxidation by manganese, is usually made in the ladle when tapping.

The Acid Open-hearth Process

We have been describing the *basic* open-hearth process, the method by which the major part of our steel is made. There is also an *acid* open-hearth process. It is largely used in the steel-castings industry. Purer raw materials must be used in this process, because no attempt is made to eliminate phosphorus or sulfur. Because of the simpler operation and care in selection of raw material, the acid open-hearth process is often regarded as being superior to the basic open-hearth process. The oxidation of silicon, manganese, and carbon are accomplished the same as in the other processes. The heat produced by the oxidation of the silicon and the manganese assists in bringing the bath to the temperature necessary for the oxidation of the carbon and helps to furnish a slag, usually quite high in FeO. Of course there is no lime boil, although a small amount of lime is often added to make the slag more fluid and to drive the FeO of the slag into the metal:

$$FeO \cdot SiO_2 + CaO = CaO \cdot SiO_2 + FeO$$

This FeO in the metal is used in carbon elimination in the final period, and a skilled operator can produce a steel with less FeO than open-hearth steel.

As to materials charged,[1] there were formerly two different procedures. One, known as the *pig-and-ore* process, used ore to eliminate the carbon from the pig; in the other, known as the *pig-and-scrap* process, only enough pig was used as would bring the carbon content up to that desired in the finished steel. The modern method, in which a charge of pig and scrap is melted, then ored down, or "pigged up," as required, may be looked upon as a combination of the two processes. Availability of materials and plant conditions modify the procedure.

THE ELECTRIC FURNACE

All the finest steels and most of the alloy steels are made in electric furnaces, because there is no contamination due to fuel

[1] J. M. Camp and C. B. Francis. "The Making, Shaping and Treating of Steel," 5th ed., p. 449, Carnegie-Illinois Steel Corporation, Pittsburgh, 1940.

and no oxidation from the furnace atmosphere. (A neutral atmosphere admits of oxidation or reduction at will). Besides, a much higher temperature can be attained, and the temperature can be easily controlled.

Fig. 10. Tapping an electric furnace. (*Courtesy Bethlehem Steel Company and Lectromelt Furnace Company.*)

Electric furnaces vary in capacity from 1 to 100 tons and obtain their heat from three electrodes, the arcs of which play on the slag of the metal bath. The furnace is circular in cross section and is constructed of a steel shell, lined with a high-quality refractory. Because there is no stack or attached fuel supply, electric furnaces are always built tilting.

Very often the steel is first melted in an open-hearth furnace and then transferred to an electric furnace for a last refining. The electric process is the only one by which the last vestige of sulfur can be removed, because there is no fuel to keep building up sulfur. Besides, as a last step after oxidation treatment, the slag can be made reducing by the addition of coke, making possible

$$FeS + CaO + C = Fe + CaS + CO$$

Coke causes the reversible reaction

$$FeS + CaO \rightleftharpoons CaS + FeO$$

to proceed to the right as desired.

As stated above, the electric furnace permits two slags to be used. The first one, the oxidizing slag, often called the *black slag,* performs the functions of the basic open-hearth slag. In the neutral atmosphere of the electric furnace, it can be changed to the reducing slag (white slag) by the addition of coke; or it can be pulled, and an entirely new white slag, consisting of lime plus a small amount of sand and coke dust, can be substituted. In either case, the desired reducing condition is indicated by the formation of calcium carbide, recognized by the smell of acetylene when it slakes in air. The electric furnace also permits of a holding period, to allow inclusions of impurities to rise to the surface.

DEFINITIONS

Steel[1] may be roughly defined as an alloy of iron and carbon that contains other impurities and sometimes alloying elements and that is ductile when cooled slowly but is rendered relatively hard and brittle when cooled rapidly. It may be differentiated

[1] E. J. Teichert, "Ferrous Metallurgy," vol. 1, p. 444, Pennsylvania State College, 1939.

from cast iron in that is contains less carbon (always under 2 per cent) and from wrought iron in that it does not contain slag streaks.

At *high temperatures:* an **acid** is the oxide of a nonmetal; a **base** is the oxide of a metal.

The term **regenerative** describes the principle of preheating furnace air and gaseous fuel by making use of checkerwork chambers which are alternately heated by hot exhaust gases and then used to preheat the incoming gas or air.

The term **reverberatory** describes the heating of a charge in a furnace by reflected heat from the roof and sides.

The term **refractory** includes any material that withstands high temperature, such as clay, sand, brick, etc. It is used for furnace linings, molds, etc.

QUESTIONS

1. What fuels are used in the open-hearth process?
2. What takes the place of fuel in the bessemer process?
3. Name the oxidizing agents used in the above processes.
4. What refractory lining is used in the basic open-hearth process?
5. Give temperatures of entering air in the blast-furnace process and in the bessemer process.
6. Give two advantages of the bessemer process.
7. Give four disadvantages of the bessemer process.
8. Distinguish between pig iron and steel.
9. Chemically speaking, what is the difference between the reactions in the blast furnace and those in the subsequent refining process?

REFERENCES

CAMP, J. M., and C. B. FRANCIS, "The Making, Shaping and Treating of Steel," 5th ed., Carnegie-Illinois Steel Corporation, Pittsburgh, 1940.

STOUGHTON, BRADLEY, "The Metallurgy of Iron and Steel," 4th ed., McGraw-Hill, New York, 1934.

TEICHERT, E. J., "Ferrous Metallurgy," 2d ed., vol II, McGraw-Hill, New York, 1944.

THUM, E. E., Editor, "Modern Steels," American Society for Metals, Cleveland, 1939.

Trans. Am. Inst. Mining Met. Engrs. Iron Steel Div., vol. 116, New York, 1935.

Chapter 3

CAST AND WROUGHT IRON

The Air Furnace

An air furnace resembles an open-hearth furnace except that in the air furnace there is not preheating of the air, since the fuel always enters at one end and the waste gases leave by a stack at the other. Thus the temperature that can be attained in the air furnace is a few hundred degrees less than that of the open hearth—too low for producing steel, but just right for producing the white cast iron that when annealed is known as *malleable iron*.

As the dimensions and design of the furnace are shown in the drawing, it is necessary only to explain the operation. (It might not be amiss to hint that the length of the furnace is related to the length of the effective flame and that the height of the furnace is usually governed by the bulk of the cold charge.)

Oil and gas have been used as fuels in the air furnace, but the almost universal practice now is to use pulverized soft coal, selected with regard to its gas, or *volatile*, content, because the gas ensures a long flame. The charge consists of scrap white iron, scrap malleable iron, scrap steel, and pig iron. The proportions are carefully calculated so as to allow for unavoidable oxidation losses during melting, unavoidable because an excess of air is necessary for good combustion. This small amount of oxidation does no harm if it is controlled and if it is uniform every day so that the charge may be calculated accordingly. An

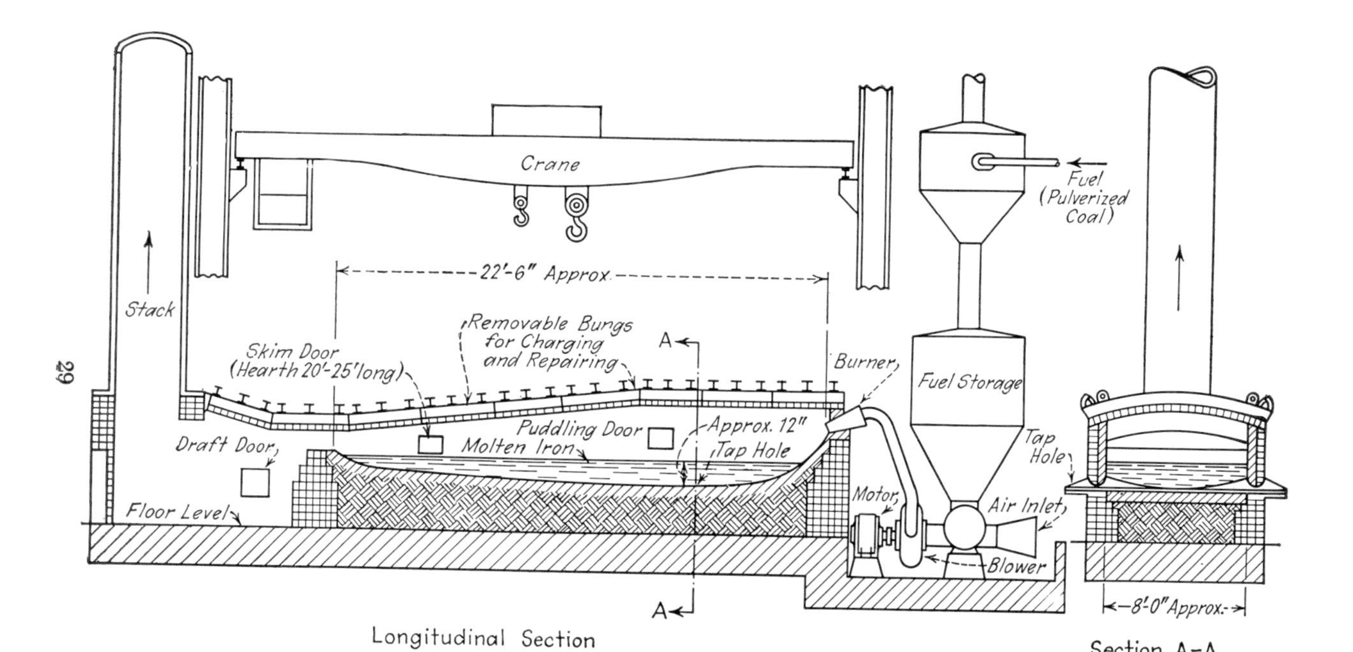

FIG. 11. Reverberatory type of air furnace used for melting malleable iron. Dimensions given for 35 to 40 tons. (*Drawn by W. G. Scarborough, courtesy of General Electric Company.*)

example of such a calculation, which is a good illustration of metallurgical calculations in general, is given on page 31.

Air-furnace Charge

White cast iron shrinks to such an extent that the *risers*, which feed hot molten iron into the casting as it is cooling in the mold, often weigh as much as the castings themselves. Risers, gates, runners, and scrapped castings, therefore, constitute about 50 per cent of the melt. This material collectively is known as *sprue* and is figured into the next day's charge according to its chemical composition.

Calculation of Charge

The first steps in calculating a new charge involve:

1. The amount of sprue available from the previous day.
2. The estimated total weight desired for the next day.
3. The analysis of the sprue that is going to be used.
4. The analysis desired for the next day.

These items determine the first item (sprue) of the tabulation on the next page.

Usually it is desired to have 1 per cent silicon and 2.50 per cent carbon in the iron as it is poured. The usual oxidation loss in melting is 0.20 per cent silicon and 0.40 per cent carbon. Hence the charge must be calculated to contain 1.20 per cent silicon and 2.90 per cent carbon.

The following materials are on hand for making up the charge, in addition to the sprue (which varies somewhat from day to day, both in quantity and in analysis):

a. High-silicon pig iron with 2.40 per cent silicon and 4.00 per cent carbon.

b. Low-silicon pig iron with 1.65 per cent silicon and 4.20 per cent carbon.

c. Some annealed malleable iron scrap with 1.00 per cent silicon and 2.20 per cent carbon.

d. Some of the latter, but rusty because bought outside the plant. The rust cuts analysis to 0.90 per cent silicon and 1.50 per cent carbon.

e. Some scrap steel, also rusty, so that both silicon and carbon can be considered zero.

For example, suppose some sprue was left from the day before, analyzing 1.00 per cent silicon and 2.50 per cent carbon and in such quantity as to constitute 50 per cent of the next day's charge.[1] The following tentative mix can then be set up:

Material	Lb. per 100 lb.	Actual lb. of silicon in the number of lb. of material listed in first column	Actual lb. carbon
Sprue	50	0.500	1.250
Malleable scrap, rusty	7	0.063	0.105
Steel	7	0.00	0.00
Low-silicon pig iron	33	0.5445	1.386
High-silicon pig iron	3	0.072	0.12
Total	100	1.1795	2.861

The amounts of silicon and carbon are now in pounds per 100 lb. and are therefore directly in percentages. It will be seen that both the silicon and carbon are short of the desired percentages, 1.20 per cent of silicon and 2.90 per cent carbon; but both elements can be adjusted to the right amount by substituting 1 lb. per 100 lb. of high-silicon pig for 1 lb. of steel. Recalculating with 6 lb. of steel and 4 lb. of high-silicon pig iron makes the totals 1.203 silicon and 2.901 carbon.[2]

Calculations such as the above are often carried out by means of simultaneous equations. Mr. Paul Green, foundry chemist

[1] It is very advisable to use all of each day's sprue for the next day, to avoid getting it mixed with other sprue not of the same composition.

[2] It is suggested that the instructor assign individual problems to each member of the class, varying the amount of available sprue from 45 to 55 per cent, the silicon content from 0.95 to 1.05 per cent, and the carbon from 2.40 to 2.53 per cent.

with the General Electric Company, suggests that the above charge can be calculated in this manner:

1. Set up the available sprue and the usual amount of malleable scrap.

Material	Lb. per 100 lb.	Silicon	Carbon
Sprue	50	0.500	1.250
Outside malleable scrap	7	0.063	0.105
Total	..	0.563	1.355

2. Subtract the amounts so found from the desired amounts, thus obtaining the amounts of carbon and silicon which must be furnished by the pig iron.

Silicon	Carbon
1.200	2.900
−0.563	−1.355
0.637	1.545

3. Equate these amounts with x, the amount of low-silicon pig iron, and y, the amount of high-silicon pig iron.

$$\text{(I)}\quad 0.0165x + 0.024y = 0.637$$
$$\text{(II)}\quad 0.042x + 0.04y = 1.545$$

4. Cancel the x's by multiplying equation (I) by 2.545.

$$\text{(I)}\quad 0.04199x + 0.0611y = 1.621$$
$$\text{(II)}\quad 0.042x + 0.04y = 1.545$$

5. Obtain value of y, $0.211y = 0.076$, and $y = 3.6$

6. Substitute value of y in equation (I).

$$\text{(I)}\quad 0.0165x + 0.0864 = 0.637$$
$$0.0165x = 0.551, \text{ and } x = 33.4$$

7. Complete the table, using these values for x and y:

Sprue	50	0.500	1.250
Outside malleable scrap	7	0.063	0.105
Steel			
Low-silicon pig	33.4	0.551	1.403
High-silicon pig	3.6	0.086	0.144
Total	94	1.20	2.902 (desired charge comp.)

It will be observed that 6 lb. of steel per 100 lb. completes the charge to 100 per cent and that this method is usually faster than the "cut-and-try" method.

Control of Reactions

About 9 hr. is required to melt 40 tons. After melting, it takes at least another hour to bring this charge up to pouring temperature and an additional hour to pour the iron into the molds. As soon as the charge is melted, a sample is taken for chemical analysis; usually half an hour later another is taken, in order not only to determine the composition when melted but also the rate of change, so that the composition may be corrected by tapping time. Additional analyses are also made at tap, in the middle, and at the end of pouring, totaling five in all. The reason for this close control will be evident later.

If the silicon is low, 50-50 ferrosilicon is added to bring it back; 80-20 ferromanganese brings the managanese content up to requirements, and a deficiency in carbon is made up with the addition of pretroleum coke (a coke with practically no ash). On the other hand, if because of reducing conditions, any of the above elements are present in excess during the melting, they must be removed by oxidation with iron ore. (Sometimes only one is high. In that case, when it is oxidized down, the other elements will be too low and must be brought back.)

These are the common control methods, but others are sometimes used. For instance, sulfur can be removed by treatment with soda ash (crude sodium carbonate), but its use is objectionable because it attacks linings of ladles or the furnace (see page 17n).

Combustion Regulation

A skillful operator is required to run the furnace. The air and coal are measured, of course, but there are too many variables to admit of running by theory. Weather conditions (temperature and moisture) affect the amount of effective oxygen per pound of air. Stack draft must be regulated so as not to pull in

FIG. 12. Number 1 melting furnace, malleable iron foundry, Erie General Electric Works. Hopper for powdered-coal feed in foreground. Stack at opposite end. (*Courtesy General Electric Company.*)

excess air. A great deal of regulation is accomplished by keeping watch of the flame; sparks indicate an oxidizing fire, while a cloudy appearance indicates too much coal and incomplete combustion indicating unnecessary fuel. The usual ratio is 750 lb. of coal per ton of iron melted.

Of great assistance in maintaining proper combustion in a furnace of this kind is some means of analyzing the waste gases that go up the stack—*flue-gas analysis.* The entering air is 21 per cent oxygen and 79 per cent nitrogen, by volume. The nitrogen performs no useful function; in fact, considerable heat is lost in raising its temperature to that of the furnace and then allowing it to pass out through the stack. Such wastes are a challenge to future engineers.

So it is the remaining 21 per cent that concerns the operator. Oxygen burns carbon to carbon dioxide, volume for volume; that is, the nearer the carbon dioxide content approaches 21 per cent, the more nearly perfect is the combustion. By what is known as an *Orsat* apparatus, the carbon dioxide, oxygen, and carbon monoxide contents of the flue gases are determined. Oxygen means excess air. Carbon monoxide means incomplete combustion. Since the Orsat analysis is somewhat time-consuming, automatic CO_2 indicators and recorders have recently come into use.

The constitution and further treatment of the product of the air furnace and the subject of white cast iron is taken up on page 40.

THE CUPOLA—GRAY CAST IRON

Gray cast iron is the product usually known as *cast iron.* Because most of the carbon is present as graphite flakes, giving it a gray color, it is called *gray cast iron.* It is as natural for graphite to form in flakes as it is for common salt to crystallize in cubes or for diamonds to be eight-sided crystals. These flakes

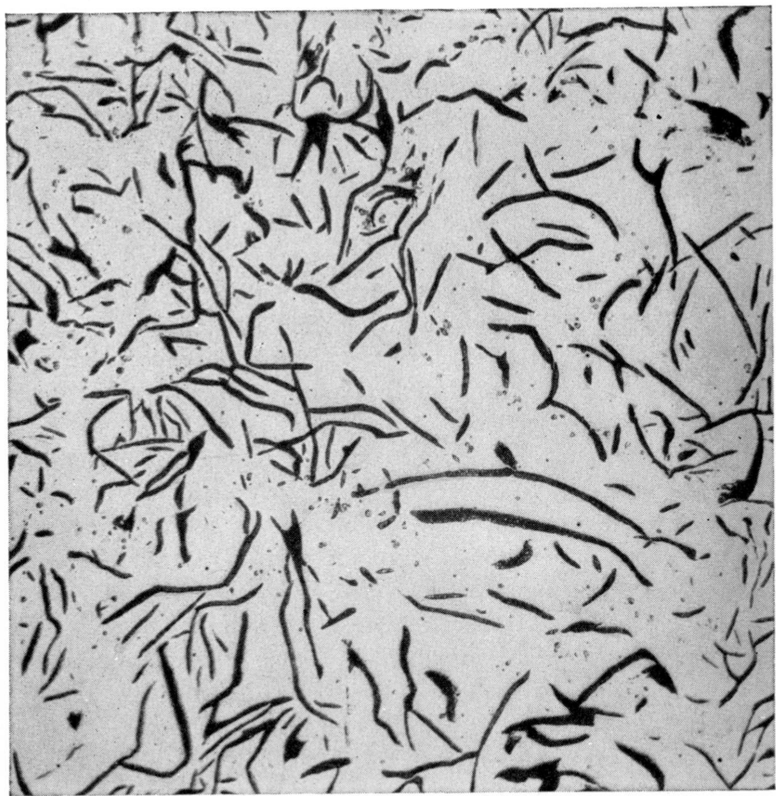

FIG. 13. Gray cast iron, polished but not etched, showing graphite flakes. Magnification 100 X. (*Courtesy General Electric Company.*)

improve machining quality but tend to weaken the iron, because of the lack of continuity between iron crystals (see Fig. 13).[1]

The sketch of the cupola (see Fig. 14) shows that it is much

[1] It must be added that gray cast iron is so much more fluid and less inclined to shrink (castings are sounder throughout) than any other iron product that it can be cast into more intricate shapes; and therefore gray cast iron has fields of usefulness which cannot be invaded, one instance being engine cylinders.

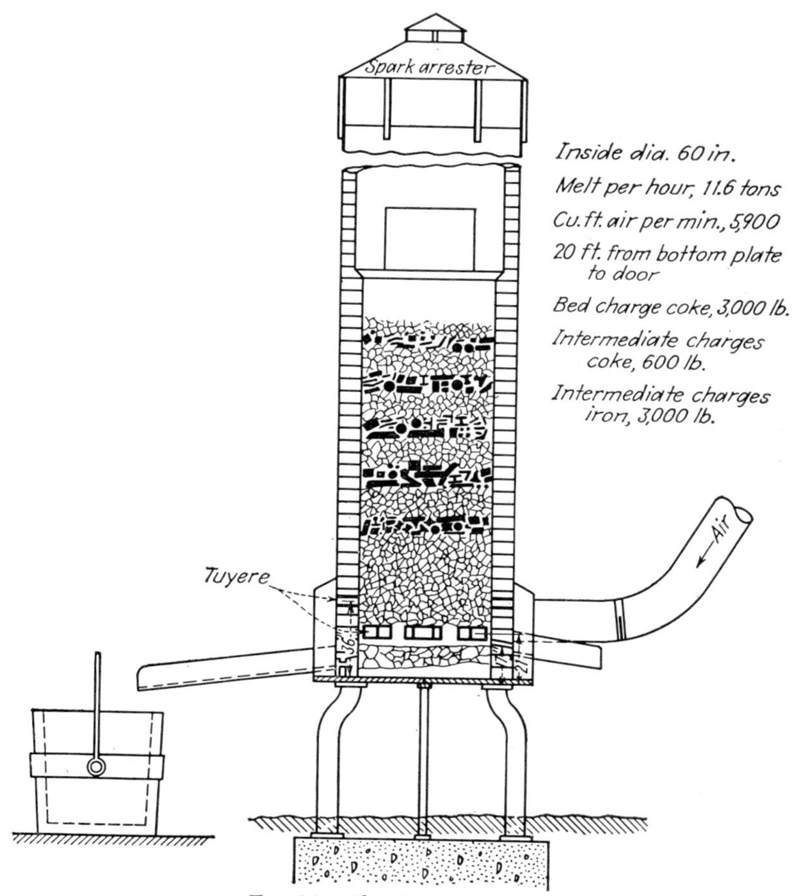

FIG. 14. Sketch of a cupola.

simpler in operation than the air furnace.[1] It is cheaper to run because, since the iron is in direct contact with the fuel, little heat is wasted. Referring to the sketch, it will be seen that alternate layers of coke and iron are charged; and, as the coke

[1] Just as in the case of the air-furnace operator, a cupola operator must be a specialist if quality iron is to be produced.

burns and the iron melts, new layers are descending into the melting zone. Thus, in a way, the operation is a continuous process. However, as linings must be repaired, cupolas are never run continuously.

Cupola Charging

The initial charge of coke placed in the bottom is measured by volume instead of by weight in order to place the first layer of iron the proper distance above the tuyeres (air ports) and thus establish the melting zone in the right location for economical and quality operation. The succeeding layers are then charged by weight, the proportion varying from 1 part of coke to 7 to 12 of iron. The iron charge consists of scrap iron, pig iron, and usually steel. It is calculated fully as carefully as the air-furnace charge and is based on requirements and operation, because little control, or "doctoring," is possible before tapping the metal. Limestone is also added to help flux the impurities such as coke ash, sand adhering to scrap, iron oxide, etc. This flux, being basic, attacks the acid lining to some extent. The high velocity of the hot gases and the wear due to scraping of the descending charge also make it necessary to repair the lining sometime during 24 hr.

Special Gray Irons

A great deal of progress has been made recently in producing cast iron of high strength, the size of the graphite particles being reduced by the addition of alloys. These alloys will be discussed in a later chapter.

There is also a trend toward what is known as *inoculation.* Late additions of some elements, for instance, in the ladle when tapping, have a different effect on the physical structure of the metal than if those elements were present in the charge as melted in the cupola.

Quoting from the 1948 edition of "Metals Handbook,"[1] "If an iron is purposely melted with a lower silicon content than desired in the casting, and a part of the silicon is added to the ladle as ferrosilicon, there is a definite improvement in structure. The mechanical properties are improved and the tendency toward chilled edges and corners is reduced."

Other *inoculants* are calcium-silicon, ferromanganese-silicon and zirconium-silicon.

An extension of the principle behind inoculation has resulted in the production of *nodular cast iron.* It can be seen that if the carbon could be precipitated in more or less spherical formations instead of in flakes, there would be much more continuity of the iron, and consequently greater strength and ductility. Such production is generally accomplished in a two-stage operation.[2] The first stage consists of the addition of a small amount of magnesium or other element tending toward the retention of carbon in the combined form. This is followed by an opposing stage in which a ferrosilicon type of inoculant, a graphite former, is added. This method of graphitization, known as *inoculation,* results in the separation of the carbon in nodules instead of flakes.

Comparison of Gray and Malleable Iron

One difference between white and gray cast iron is that the latter contains more carbon; as you have seen, it was produced in contact with the fuel, coke. That is not, however, the principal difference. The fact that in white cast iron all the carbon is in combined form is of much greater importance. In white iron none of the carbon exists as free carbon. Even the best microscope cannot find a particle of free carbon, because

[1] Page 515.

[2] Articles by G. Vennerholm, H. Bogart, and R. Melmoth, in *Materials and Methods,* April, 1950, p. 51. Published by the Ford Motor Company.

this element is in combination with the iron as the compound Fe_3C (cementite), one of the hardest substances known (see description, pages 69 and 79). It is as brittle as glass and is therefore of very little use in this condition. But when it is subjected to a conversion process by controlled temperatures (commonly known as *malleabilizing*), it becomes a product of great ductility, strength, and machinability.

CONVERSION OF WHITE CAST IRON TO MALLEABLE IRON

As will be explained in Chap. 5, iron can hold much more carbon in solution at high temperatures than at lower temperatures. The more the solution cools, however, the more tendency there is for the carbon to separate. This tendency is counterbalanced by increasing rigidity: therefore, it is easy to obtain a *supercooled*[1] solution of carbon in iron, which will be considered in greater detail under *heat-treating*, and we can see that the cementite thus formed is not an extremely *stable* substance.

Under the proper conditions the carbon can be caused to separate out as free carbon. These conditions are temperature, time, and the presence of a class of elements, such as *silicon*, which are known as *graphitizers*. Of course, the percentage of carbon has a great influence also. There is such an excess in gray cast iron (3.50 per cent, as compared with 2.50 per cent in white iron) that when it is combined with a greater percentage of silicon (2 per cent, as compared with 1 per cent in white iron) the result is the separation of carbon into graphite flakes in the case of gray iron, as has been previously mentioned.

The composition of white iron has been purposely controlled

[1] It is possible for liquid solutions to be *supercooled*, meaning that precipitation or freezing can, under certain conditions, be delayed so as to start at a lower temperature than the normal one.

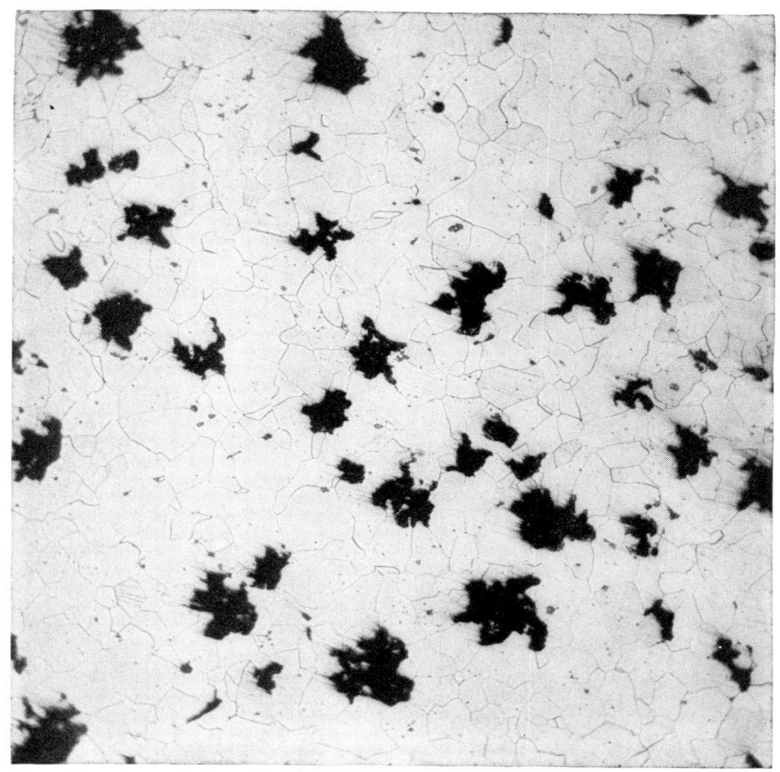

FIG. 15. Annealed malleable iron at low magnification (60 diameters), etched, showing ferrite grains with nodules of carbon. (*Courtesy General Electric Company.*)

with the view of preventing such precipitation of graphite on its original cooling. The first step in converting brittle white-iron castings into ductile, shockproof malleable iron consists in heating the castings to a temperature that puts the carbon into solution with a certain freedom to migrate. To prevent warping and softening of the castings, this temperature should not greatly exceed 1800°F.

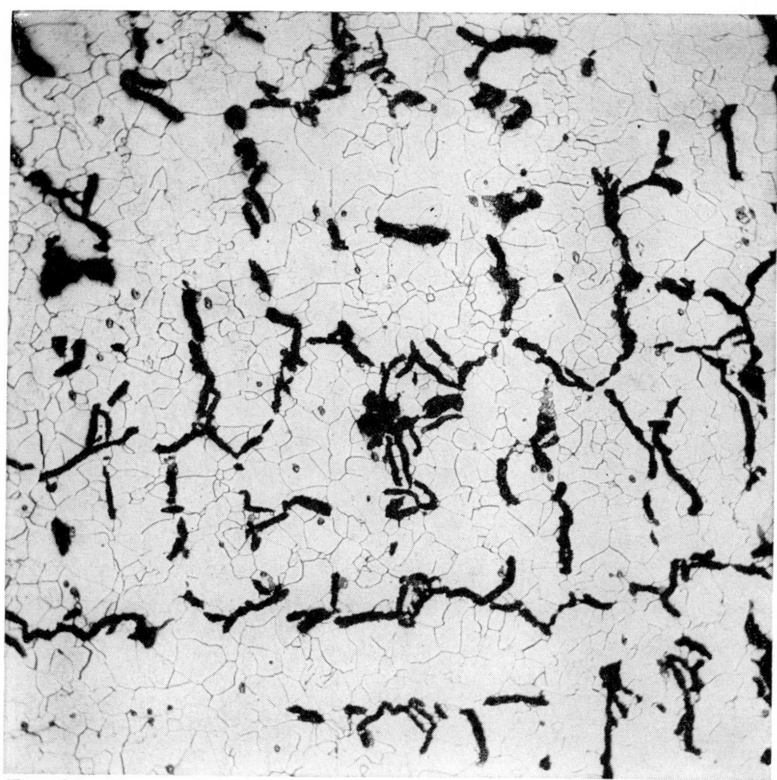

FIG. 16. Poor malleable-iron structure due to excess of silicon or carbon. Notice lack of continuity in ferrite matrix. Magnification 100 diameters. (*Courtesy General Electric Company.*)

After the castings have soaked at this temperature until the entire load is uniformly heated, the temperature is lowered gradually. This treatment causes the cementite to break up, and its carbon begins to separate. It cannot assume its natural state, flake graphite, because of the solidity of the surrounding iron. Instead it separates into smaller spherical nodules, which do not destroy the continuity of the iron background or *matrix.*

This characteristic is well illustrated by a comparison of structures shown by photomicrographs (see Figs. 13 and 15).

The conversion of combined carbon is not completed in the initial breaking up of cementite mentioned above. As noted in Chap. 6, some of the cementite will still remain, interspersed as plates, or layers, of alternate ferrite[1] and cementite. This is known as *pearlite* and is considerably more stable than cementite.

FIG. 17. Three-hundred-and-fifty-kilowatt electric annealing furnaces, short-cycle process (24 hr.), Malleable Iron Foundry, Building 22, Erie Works. (*Courtesy General Electric Company.*)

The conversion of this remaining pearlite takes about three times as long as the initial breaking up of cementite, but it is accomplished at a lower temperature. (The breaking up of the pearlite would not be possible if there were not carbon present already—as nodules.)

[1] *Ferrite* is a name given to carbonless iron, which, however, may contain small amounts of silicon or other alloys. Ferrite is explained more fully in Chap. 5.

It can now be realized why such close control of composition is necessary: Too much carbon would result in flake formation when cast (see photomicrograph, Fig. 16); too small an amount would result in not enough cementite to ensure a separation of carbon. The percentage of silicon is just as important. It assists in drawing the carbon out of combination. Too large an amount, about 1.10 per cent, would result in flake formation (*primary graphite,* as it is called—primary because it was formed previous to the conversion process). Silicon is known as a *graphitizer; i.e.,* it assists carbon to separate from the Fe_3C carbide form.

The Navy specifications for malleable iron call for 53,000 p.s.i. (pounds per square inch) and 18 per cent elongation. By a modification in the annealing cycle or in composition, a product known as *pearilitic malleable iron* can be produced. This has a tensile strength of 60,000 to 70,000 p.s.i. without a great deal of decrease in elongation. As the name implies, it consists of carbon nodules in a pearlite instead of a ferrite matrix.

WROUGHT IRON

A century ago, before the bessemer or open-hearth processes of steelmaking were known, all steam boilers, rails, bridges—all iron plates and rolled shapes—were made of wrought iron; and much of this old iron still survives, for example, in ornamental iron work. In fact, because of its excellent corrosion resistance, wrought iron is still used for such work.

Although comparatively only a small quantity is now made, it possesses qualities which keep it in demand. It is a unique ferrous product in more ways than one. To begin with, all other iron or steelmaking processes finish with the metal in a molten condition, while wrought iron is produced in a spongy or plastic state.

Reactions

The solubility of many impurities in iron increases with temperature. This is especially true of gases. Thus, as the iron descends toward the hearth in the blast furnace and increases in fluidity (both because of increase in temperature and of increase in carbon content), it takes up several impurities. The "irony" of this situation is that these impurities must be removed later and at considerable expense.[1]

Now, if we could operate this process that is going on in the lower half of the blast furnace *in reverse,* starting out with molten pig iron in the hearth and taking it back up to pure spongy iron by shedding the impurities it acquired on the way down, the result would be a very pure iron.

In principle, that is exactly the way wrought iron is produced at the present time. It is accomplished by one of two methods. By the old-fashioned method, known as the *puddling* process, a few hundred pounds of pig iron is melted in a furnace resembling an air furnace in that it cannot reach the 3000°F. necessary to melt pure iron, and it is open enough on one side to permit handworking of the charge.

The melting pig iron is packed in iron oxide, which makes a very basic slag that will take up silicon, phosphorus, and sulfur. Iron oxide does not enter into the iron because the carbon in the iron is protection against it. Besides, the iron is *shedding* oxides. Their solubility decreases as the iron approaches solidification. However, it is necessary to bring the slag into intimate contact

[1] Impurities acquired by the iron because they have been reduced simultaneously with the iron. These are then removed in steel-making processes by oxidation. Then, after all this, the oxygen must be removed (see deoxidizers, page 14).

For many years sponge iron has been made by direct reduction experimentally and in a few small plants. To date such a process has not been able to compete commercially with the blast furnace.

with every particle of the stiffening iron so that the impurities can enter the slag. This working, *rabbling* as it is called, must be done by hand, which is a very hot and arduous job, and is the reason that only a few hundred pounds can be made in one "batch." When the ball of iron with its intermixed slag has reached a plastic stage, it is removed and run through a *squeezer* to remove the slag. About 1 to 3 per cent of the slag is retained, giving wrought iron its characteristic fibrous appearance when these slag particles are elongated by squeezing or rolling.

The Aston Process

In 1927, the A. M. Byers Company of Pittsburgh instituted what is known as the *Aston* process for quantity production of wrought iron. Pig iron is melted in a cupola, poured into a bessemer converter, and refined bessemer steel produced in 10-ton batches.

Meanwhile a slag is being prepared, the ingredients of which are iron ore, sand, roll scale, and leftover slag. The aim is to simulate the ferrous silicate slag produced in the puddling process and whose melting point is much lower than that of the refined iron.

This slag is prepared and melted in two basic-lined, oil-fired, tilting, open-hearth furnaces and poured into a large receptacle called a *thimble*. The ladle containing the steel is now placed in position and the steel poured into the slag, being constantly moved so as to distribute it evenly as it pours through. Essentially the steel is quenched, for the temperature of the slag is much lower than the solidifying point of the steel. The steel forms into small shot as it is being poured. In fact, the operation is known as *shotting*. Because of the release of gas in the pellets, they actually explode; thus a great amount of surface is exposed to the purifying action of the slag.

The metal collects in the bottom of the thimble and, being of

FIG. 18. Pouring the molten refined iron into the liquid slag. (*Courtesy A. M. Byers Company.*)

good welding temperature, forms into a mass—which of course is thoroughly impregnated with slag. The slag is poured off, and the ball of metal is squeezed in a hydraulic press, forming an ingot. Without reheating, this is rolled as desired—a great deal of it rolled into skelp for wrought-iron butt or lap-welded pipe.

Properties of Wrought Iron

Besides its corrosion-resistant properties, wrought iron is noted for its shock and fatigue resistance. This property can be determined by what is known as a *notch test,* briefly described as follows: Suppose we put a bar of steel in tension and file a small notch on one side. The pulling force will be distributed evenly in the cross section, but at the notch the strain lines will be

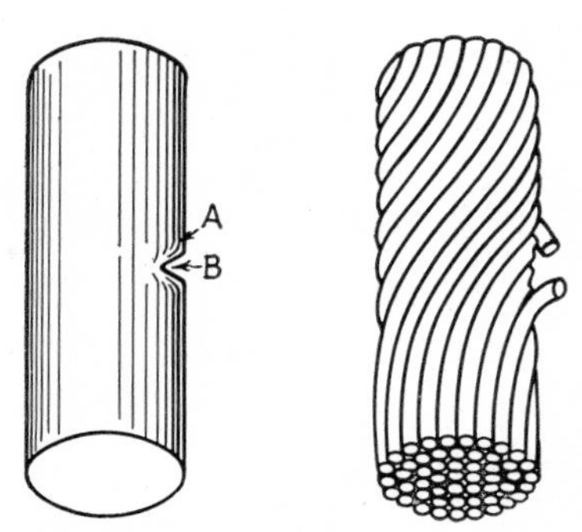

Fig. 19. A few strands of a cable may be broken without weakening the cable very much, but cut a notch in a solid steel bar and it will fail because of concentration of force at *B*.

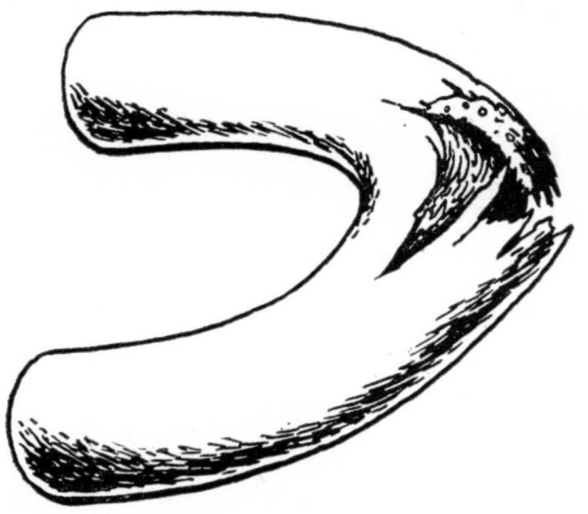

Fig. 20. The illustration will give you a good idea of the fibrous structure of wrought iron. Note that it breaks almost like a piece of tough, resilient hickory. (*Courtesy A. M. Byers Company.*)

deflected inward (*A*, in Fig. 19), causing a concentration at *B* which can rapidly develop into a failure.

On the other hand, consider a cable composed of small wires. If a few strands are cut, corresponding to the filed notch in the solid bar, there will be no deflection inward, and concentration of strain is avoided. Because of its fibrous structure, a strained rod of wrought iron would resemble a strained cable more than a strained solid-steel rod.

DEFINITIONS[1]

Cast irons are alloys of iron containing so much carbon that, as cast, they usually are not appreciably malleable at any temperature.

Wrought iron is a ferrous material aggregated from a solidifying mass of pasty particles of highly refined iron with which, without subsequent fusion, a minutely and uniformly distributed quantity of slag is incorporated.

Cementite is a hard, brittle, crystalline compound, represented by the formula Fe_3C.

By **matrix** is meant the ground mass, or principal substance, in which a constituent is embedded.

Malleable iron is white cast iron converted by heat-treatment into a malleable product consisting of a ferrite matrix that contains nodules of carbon.

QUESTIONS

1. Why is it necessary to ensure careful chemical control of elements in white cast iron?
2. What element besides carbon promotes the breaking up of iron carbides into carbon and iron?
3. Explain the effect in the furnace of rust on material charged.
4. Why is white cast iron hard and brittle as cast, but soft and malleable after annealing?
5. Why is carbon in the form of large flakes in gray cast iron, but formed into nodules in malleable iron?
6. What oxidizing agent is used in the puddling process?
7. Name the two forms of Fe_3C that are broken up (transformed) in the malleable-iron-annealing furnace.
8. Why does cupola melting result in a greater percentage of carbon than air-furnace melting?
9. Explain the resilience of wrought iron.

[1] TEICHERT, op. cit., p. 419.

REFERENCES

American Foundrymen's Society, "Alloy Cast Irons," Chicago, 1939.

Aston, James, and Edward B. Story, "Wrought Iron," A. M. Byers Company, Pittsburgh, 1949.

Donoho, C. K., Producing Nodular Graphite with Magnesium, *Am. Foundryman,* Chicago, February, 1949.

Eash, J. T., Effect of Ladle Inoculation on the Solidification of Gray Iron, *Trans. Am. Foundrymen's Assoc.*, vol. 49, Chicago, 1941, pp. 887–910.

Malleable Founder's Society, "American Malleable Iron," Cleveland, 1944.

Merica, Paul D., Progress in Improvement of Cast Irons, *Trans. Am. Inst. Mining Met. Engrs., Iron Steel Div.*, vol. 125, New York, 1937, p. 13.

Mulcahy, B. P., Blast Variations Seriously Affect Cupola Operations, *Am. Foundryman,* Chicago, June, 1948.

Swartz, H. A., "American Malleable Iron," Penton, Cleveland, 1922.

Chapter 4
CONSTITUTION DIAGRAMS OF ALLOYS

The definition of an alloy, to be comprehensive, can be stated as "a solution of two or more elements, at least one of which is a metal, or the mixture of metallic crystals resulting from solidification or cooling of such a solution."

Effect of Alloying

Alloying to some extent is so common that pure metals are the exception rather than the rule. Sometimes the metal contains small amounts of alloying impurities for the reason that it is such a difficult matter to obtain it pure, but more often the alloying has been done with definite properties in view.

Often the addition of minute amounts of one element to a metal has a profound effect on its properties. For instance, pure iron has a tensile strength of about 40,000 p.s.i. But absolutely pure iron is very difficult to obtain. Iron that is commercially considered quite pure stands 50,000 p.s.i. tensile strength. Now, the purposeful addition of less than 1 per cent of carbon to this iron (0.80 per cent, to be exact) raises its tensile strength to 125,000 p.s.i. without any attempt to take advantage of "heat-treat." With such heat-treatment, its tensile strength can be increased to 180,000 p.s.i. This control of properties in connection with iron is taken up in the next chapter. In fact, the present chapter is a preparation for the next.

Considering the magnitude of its effect, the 0.80 per cent mentioned above is a small amount, but there are many instances of just as great an effect with smaller percentages of alloy. As little as 0.02 per cent of phosphorus reduces the conductivity of copper 20 per cent, and 0.1 per cent reduces it 26 per cent.

Salt and Water System

These additions of one metal or element to another affect the melting and freezing characteristics of the metals. Pure metals freeze or melt at definite temperatures, just as pure water does.

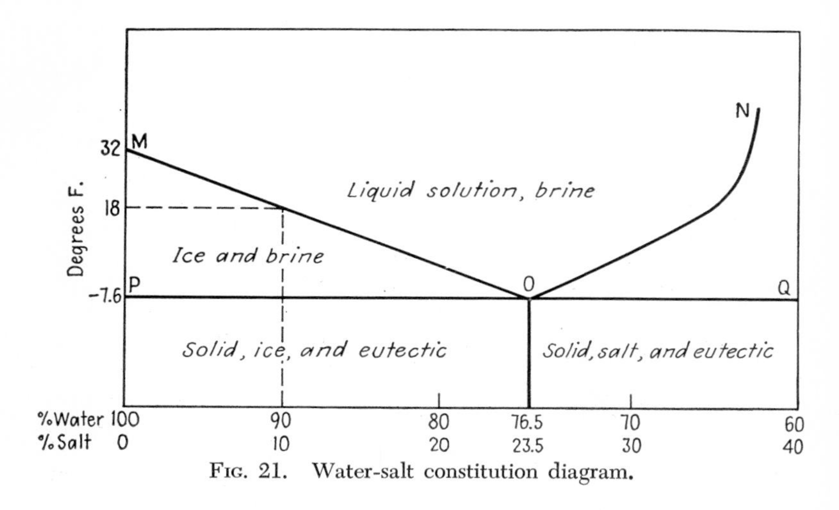

FIG. 21. Water-salt constitution diagram.

Under ordinary conditions of pressure, water freezes or ice melts at 32°F.; pure gold melts at 1943°F.; pure lead at 621°F.; etc. But add some common salt to the water, and a lowering of the freezing point results. This is a matter of common knowledge to everyone, for example, the use of salt on icy sidewalks to thaw the ice. Figure 21 is a constitution diagram of the water-salt system.

Notice that the diagram is plotted vertically in temperature and that it is plotted horizontally in "percentage composition," meaning that it begins at the left with 100 per cent water. That, of course, means pure water with 0 per cent salt; but as one moves to the right, the figures show progressively more salt to less water, that is, a stronger solution of salt (commonly called *brine*). Above the line *MON*, the solution is entirely liquid. Below the line *POQ*, it has solidified completely. Let us study what the diagram tells us. Pure water freezes at 32°F., but add 10 per cent salt and the point where the solution starts to freeze is lowered to the 18°F. temperature line (dotted). Note also that the vertical dotted line indicates a 10 per cent salt solution. This 10 per cent solution does not freeze entirely at 18°F.; it merely begins there. Then, as it cools down toward –7.6°F., it is in a mushy or granular condition, not frozen completely until the temperature of –7.6°F. is reached. What are these granules? They are pure ice. As they crystallize out from the liquid, they leave the liquid more and more salty until at —7.6°F. the brine, which was orginally a 10 per cent solution, has now become a 23.5 per cent solution from the loss of water freezing out as ice. In fact, it is a 23.5 per cent solution which always freezes at —7.6°F. If the solution originally had been 23.5 per cent salt, it would not have frozen at all until this low temperature was reached. That strength solution is known as a *eutectic*, being the per cent solution having the lowest freezing point of any system or series of mixtures.

Construction of Constitution Diagrams

Now, a few words as to how constitution diagrams are derived. If either a solid or liquid is heated considerably above the temperature of its surroundings and allowed to cool again without going through any change of state, *e.g.*, not changing from liquid to solid, and its temperature is recorded continually so that it may be plotted against time, the result would be a regular curve (see Fig. 22).

Because at the beginning of cooling there is the greatest difference in temperature between the material and its surroundings, the curve slopes steeply at first; *i.e.*, it cools rapidly per unit of time. Of course, when nearing the temperature of its sur-

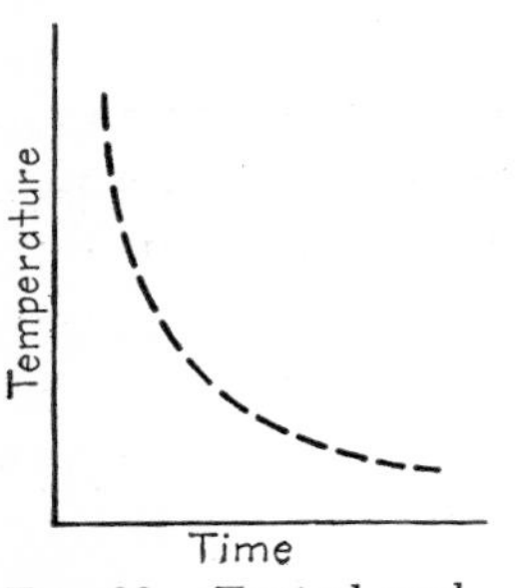

FIG. 22. Typical cooling curve where there is no change of state.

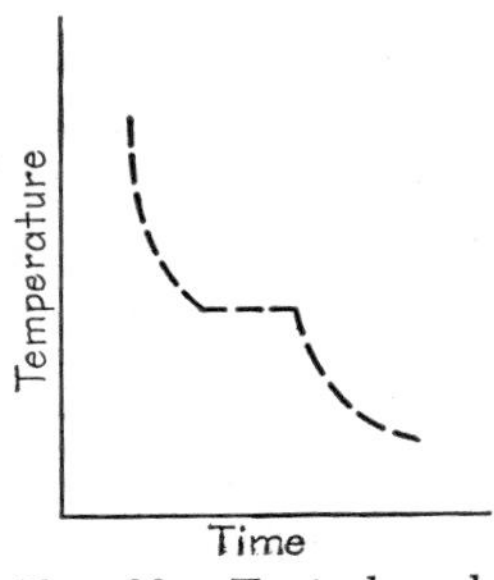

FIG. 23. Typical cooling curve where the material goes through a change of state—such as liquid to solid—at constant temperature.

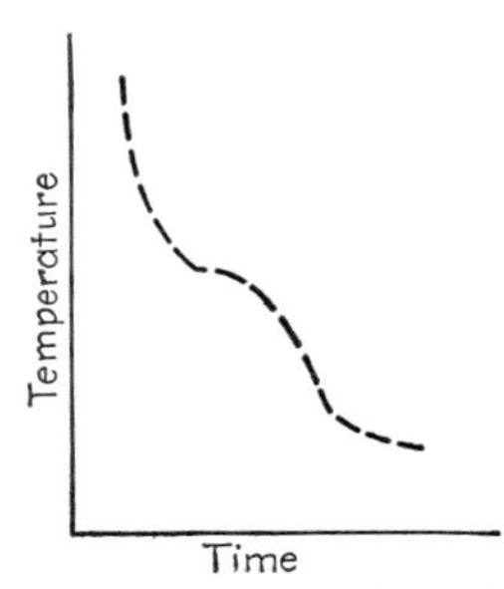

FIG. 24. Typical cooling curve of an alloy which solidifies progressively over a range of temperature, instead of at constant temperature as would a pure metal.

roundings, its rate of cooling is much slower, but the point to be emphasized is that the line curves in a regular manner throughout its length.

If, however, a change of state occurred during the cooling, say, from liquid to solid, as in the case of the freezing of water, the curve would be regular down to the freezing point. Then, because of the giving up of latent heat of fusion, the line would continue horizontally for a certain time until the freezing was completed. After that, the curve for the cooling of the solid would continue on in its regular manner. This kind of freezing curve is typical of pure water, pure metals, and other stable compounds (salts, for example) and of that type of alloy called *eutectic*.

Solutions of one metal in another do not present such simple cooling curves throughout their range of percentage composition. The character of their cooling curves depends greatly upon their solubility in the solid state. Many pairs of metals are soluble in each other in the liquid state. Not all of these are mutually soluble in the solid state. Such variations would cause variations in their cooling curves.

Even in the case of metals that are mutually soluble in all proportions in both solid[1] and liquid phases (which we shall call Case I), the freezing of the mixture never occurs at one certain temperature but extends over a range, in which cooling is retarded but not stopped. The cooling curve for such an alloy would appear as in Fig. 24.

The alloys of gold and silver present an example of Case I. Figure 25 shows a diagram constructed to show all possible alloys and all possible percentage compositions, beginning with 100 per cent gold and 0 per cent silver, at the left, and extending through the entire range of compositions to 100 per cent silver and 0 per cent gold, at the right.

The dotted lines in the diagram (see Fig. 25) represent some of the cooling curves from which a constitution diagram would be formed. Notice that the dotted lines at the extreme right and left indicate a pause in the normal cooling rate. Temperature remains constant at that point while time goes on (horizontal part of the dotted line). This indicates a change from liquid to solid at *constant temperature*, characteristic of pure metals and eutectics.

[1] This brings in the term *solid solution*, which may seem a trifle odd to anyone accustomed to considering a solution to be a liquid. A solution may be a solid as long as it satisfies the definition "a homogeneous mixture whose proportions can vary within wide limits." Glass is an example of a solid solution.

However at intermediate points, the dotted lines do not become horizontal; the freezing does not occur at one temperature but is distributed over a range of temperature. In such cases the diagrams are constructed from the points on the cooling curve where there is a *change in rate of cooling.* After a large number of such points have been determined, the smooth curves *o* and *p* are drawn, *o* representing the beginning of solidification on

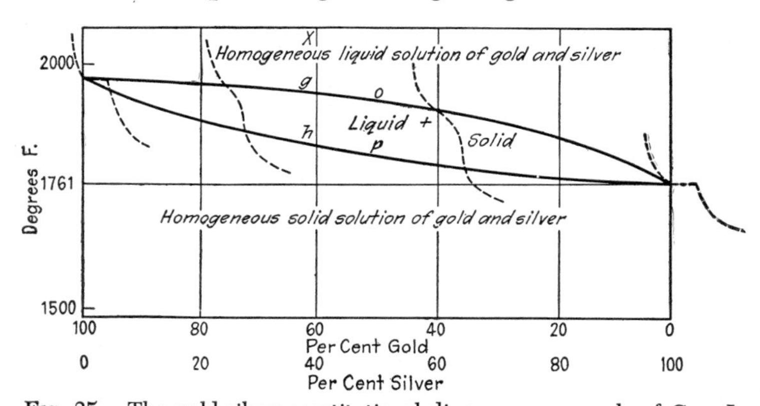

FIG. 25. The gold-silver constitutional diagram an example of Case I.

cooling and *p* representing the completion of solidification. The line through *o* is called the *liquidus,* and the line through *p* is called the *solidus.*

Case I

Any alloy of the gold-silver series going through the temperature range between *o* and *p* will progress through a *mushy stage.* As an example, let us consider an alloy of composition *X.* When cooling from a melted condition, it will remain liquid until it reaches the temperature indicated by *g.* At that point a few crystals begin to solidify. More of the liquid will solidify continually until the solidus is reached at *h,* when the mass will be entirely solid. It will be a solid solution of one metal in the other so uniform and homogeneous that no microscope could differentiate gold or silver crystals. Other binary alloys that behave similiarly to the gold-silver series, because they are soluble in all proportions, liquid or solid, are gold-platinum, iron-nickel iron-manganese, and nickel-chromium.

These pairs of metals are mutually soluble in each other in the solid state because their crystal structure is so similar. Of course the crystals are not identical, or there would be no hesitation and no mushy stage, in freezing or in melting.

Most pairs of metals differ enough in their crystalline form to present quite a difficulty of arrangement in solidifying together. For instance, it would be quite a problem in solid geometry to build up a solid out of cubes and pyramids without voids or strains. This same explanation applies to melting as well as to freezing. If the atoms were forced together in a strained and unnatural formation, those very strains would act to pull the atoms apart and thus lower the melting point. Eutectics are never homogeneous solutions. They usually solidify in laminations, and their constituent parts are visible under the microscope.

Case II

Such is the situation in Case II. Two metals are mutually soluble in their liquid state, but they are insoluble in their solid state. Figure 26 shows a cooling curve characteristic of such an alloy. This figure shows that the *rate* of cooling is lowered during the range of temperature where solidification of one constituent is in progress. (This much corresponds to Case I.) But notice that the curve flattens out (b to c) while the residual eutectic is solidifying at constant temperature. This horizontal line is longest for compositions of the eutectic ratio (point E, Fig. 27), because the composition there is 100 per cent eutectic; and it gets shorter toward the left, where of course it disappears when it reaches the pure metal line. (It is then a horizontal line at the melting temperature of the metal.)

Figure 27 shows a general example of a Case II diagram. The addition of some metal B to A lowers A's freezing point, and this freezing point is progressively lowered as the percentage of B increases. The same thing is true on the right side of the diagram with regard to the addition of A to B. Naturally, these two descending curves intersect at a point E; this lowest melting point is known as the *eutectic point*. The term *eutectic* is applied to the alloy or solution having the lowest melting point possible of a given series of compounds. See the end of the chapter for another definition.

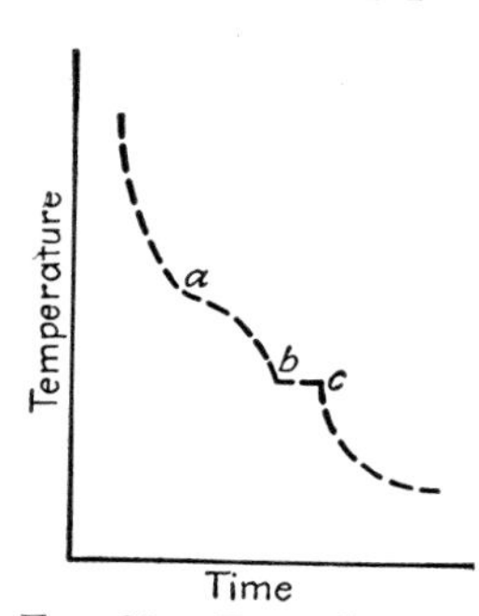

FIG. 26. Typical cooling curve of an alloy which is soluble in its liquid state, but which because of insolubility in the solid state first deposits crystals of one phase in a "mushy stage" of freezing (a to b), and finally the remainder solidifies completely at a constant temperature (b to c) while crystals of a "eutectic" are forming.

The diagram is worthy of study in several respects. As the line AEB was determined from points where solidification began for the percentage composition indicated, it is evident that it is natural for the composition to be just on the point of starting to freeze when cooling or just on the point of being completely melted when heating; we say that it is *in equilibrium*.

This word *equilibrium* is important in this study. Note the cooling of an alloy of 40 per cent B and 60 per cent A in the diagram (see Fig. 27); when it had cooled to the point W, it could have stopped as soon as solidification had just started and a seesaw begun, by causing the solidified part to melt or by causing solidification to begin again, according to whether the temperature was raised or lowered a degree or two.

This idea of balance is of great value in understanding constitution diagrams.[1] No matter how much solid is present, any point on the line *VW* would have the same composition in the liquid phase. Take the case of cooling a mixture of 20 per cent *B* and 80 per cent *A*. There is a greater percentage of *A*, and crystals of *A* would begin to solidify at *X*. On the cooling of

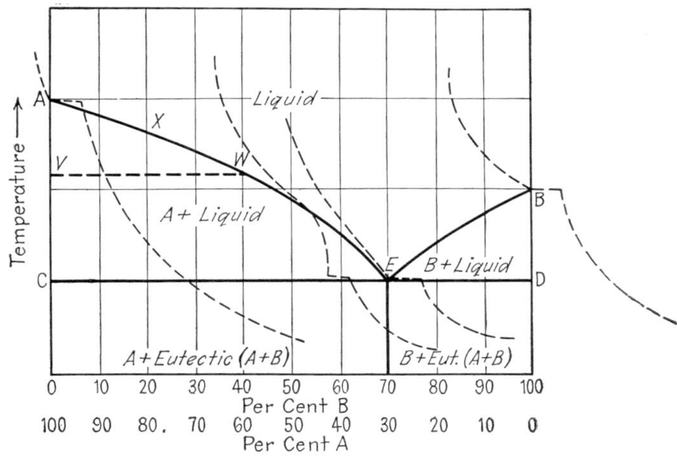

FIG. 27. General example of Case II.

this 20–80 alloy to the line *VW*, just enough *A* crystals would have solidified so that the liquid composition at the line *VW* would be 40 per cent *B* and 60 per cent *A*. It could not be any other way, because such is the liquid composition that is in balance, or in equilibrium, at that temperature. This fact is so important that it will be reviewed in connection with the cadmium-bismuth diagram.

For convenience, this principle (that the same composition of liquid prevails along any horizontal temperature line in any separate *block* of a constitution diagram) will be referred to as

[1] In fact, constitution diagrams are often referred to as *equilibrium diagrams*.

the *principle of horizontals.* In applying this principle to the line *CED*, one sees that the liquid phase touches this line at *E*, which has the composition, 70 per cent *B* and 30 per cent *A*; hence all liquid freezing (or all solid melting) anywhere along this line is of that composition. This is *eutectic* composition; and as eutectics freeze or melt at a constant temperature (as pure metals do), it is immediately seen why all of the cooling curves show a straight horizontal line at this temperature. This horizontal line is longest at *E* because that point has 100 per cent eutectic, and it decreases in length toward *C* or *D*, where of course it vanishes.

Case III

Before the study of a practical example of Case II, another case (Case III), one that appears in most of the nonferrous-alloy diagrams of Chap. 11, must be explained.

The general case is sketched in Fig. 28. It resembles Case II (see Fig. 27) as far as the liquidus lines and eutectic point are concerned. The added complexity consists of the lines *AFM* and *BGN*. They are there because the two metals are *partly*[1] soluble in each other in the solid state.

Perhaps a review of the three cases will be helpful at this time: Case I is soluble in both the molten and solid states; Case II is soluble in the molten state and insoluble when solid; Case III is soluble in the molten state and partially soluble when solid.

The changes that occur in Case III to the left and above the point *F*, on cooling, greatly resemble Case I. There is a solid solution of metal *B* in metal *A*.[2] But there is quite a difference

[1] Not only *partly* but in almost all cases *variably* soluble according to temperature.

[2] The customary designation for the first solid solution at the left of a series is the Greek letter α (alpha); the second, to the right, is called β (beta).

below F. As the temperature is lowered, the solubility of B in A decreases; thus, even though the solution is solid, a change occurs (B separates out, usually in the form of β solution). This change is represented by line FM. Corresponding changes take place on the right of the diagram.

Applying the principle of horizontals, we can tell that although in Case II the eutectic was a mixture of metal A and

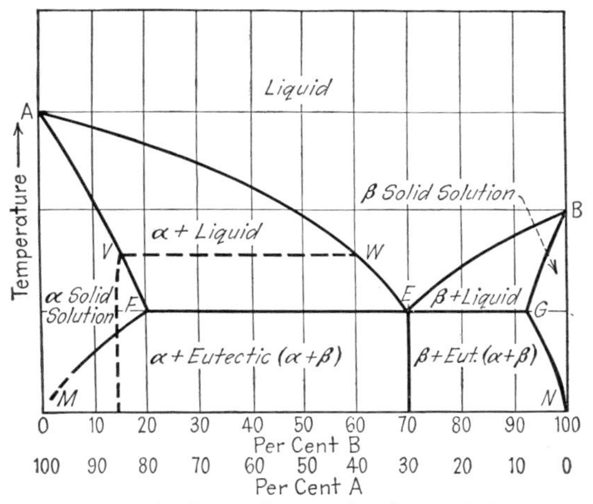

FIG. 28. General example of Case III.

metal B, because the lines EC and ED (see Fig. 27) extended out to the extreme edge of the diagram—*to the line of pure metal*—in this case the eutectic is a mixture of *saturated* α and β solutions, because the corresponding lines in Fig. 28 end at F and G, and these saturated solutions have compositions as indicated by dropping vertical lines.

Also let us apply the principle to calculate the character and percentage of constituents in the mushy stage AFE in Case III. Anywhere along the line VW the liquid part has the composition as indicated by dropping a vertical line, just as in Case II. But there is a difference with regard to what has solidified down to *VW*, for the point *V* rests on a line indicating solid α solution instead of pure metal as in Case II, and this solid solution has the composition indicated by its vertical, 15 per cent *B* and 85 per cent *A*. The lead-tin system is a good example of Case III (see Fig. 111, page 205 for diagram and discussion).

Case IV

Case IV comprises solubility in the molten state and, for a certain temperature range, solubility in the solid state. In this respect it resembles Case I. As the solid solution reaches a certain

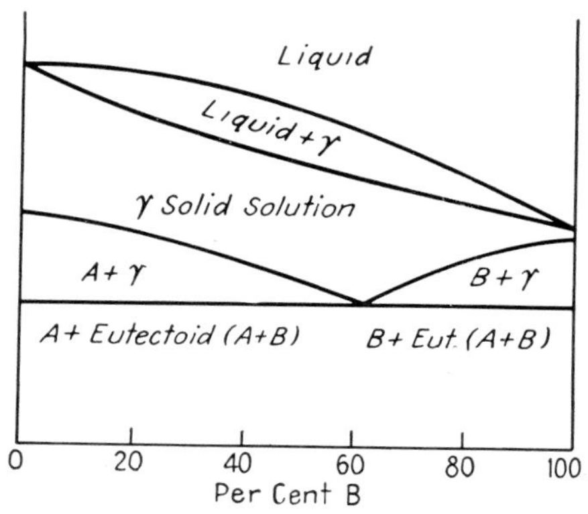

Fig. 29. General example of Case IV.

lower temperature, however, it decomposes into two new solid phases. As shown in Fig. 29, the transformation resembles a eutectic reaction except for the fact that it occurs within a solid instead of marking a transition from liquid to solid which characterizes the pure eutectic reaction. (The suffix *oid* signifies *resembling*. See definition at end of chapter.) Case IV resembles the iron-iron carbide diagram that will be taken up in the next chapter. Additional cases will be discussed in Chap. 11 in connection with nonferrous metals.

Problems

In order to familiarize the student with some of the characteristics as outlined by constitution diagrams, it is suggested that the instructor assign problems on the quantitative relations of constituents at different points on a diagram.

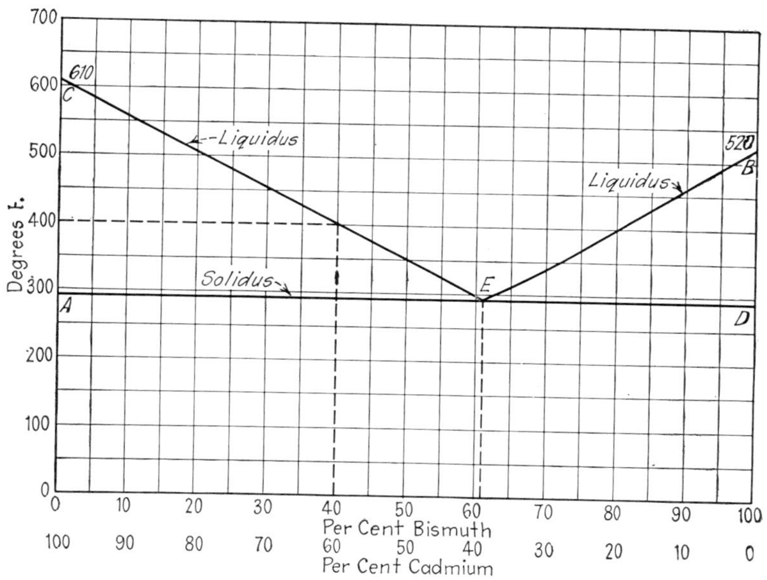

FIG. 30. The bismuth-cadmium constitutional diagram.

Figure 30 shows the bismuth-cadmium system, which has been chosen as a basis for these problems because of its simplicity. In figuring compositions during and after solidification, only two principles need be kept in mind:

1. The composition of the liquid portion is the same along any horizontal line within the triangles *CAE* or *BDE* and is equal to the composition indicated at the point of intersection of the horizontal line with the liquidus. This composition is read on

the percentage scale below the diagram by dropping a vertical line from the point of intersection to the percentage scale.

2. The application of principle 1 to the 295°F. temperature line means that any metal freezing at that temperature must be of eutectic composition (61 per cent bismuth and 39 per cent cadmium). Furthermore, no bismuth freezes above this line in alloys to the left of the eutectic point, and no cadmium freezes above this line in alloys represented to the right of the eutectic point.

For the purpose of illustration, let us consider a few specific cases. First, let us follow the cooling of an alloy of 80 per cent cadmium and 20 per cent bismuth. It is in the molten state above the liquidus *CE*. Upon reaching this line, cadmium starts to solidify out of the melt. As it cools between the liquidus and solidus lines, the liquid portion becomes progressively richer in bismuth because of cadmium crystallizing. When the solidus is reached, the remaining liquid consists of the original 20 per cent bismuth plus enough cadmium to satisfy the eutectic proportion of 61 parts of bismuth to 39 parts of cadmium. For instance, if 100 lb. of the 80-20 alloy is used for this problem, the amount of the metal (eutectic) freezing at 295°F. must satisfy the proportion.

$$\begin{aligned} &\mathrm{Cd}:\mathrm{Bi}::\mathrm{Cd}:\mathrm{Bi} \\ &x:20::39::61 \end{aligned}$$

where $x =$ pounds of cadmium $= 12.8$, and the total metal freezing is $12.8 + 20 = 32.8$ lb. (The remainder of the cadmium, 70 minus 12.8, or 67.2 lb., was frozen out as the alloy cooled between the liquidus and solidus lines.)[1]

Next, let us compute the amount of cadmium that has crystal-

[1] In this case cadmium is known as the *excess constituent*, and that part of the diagram to the left of the eutectic point is spoken of as the *cadmium-rich* side.

lized (solidified) at some point between the liquidus and solidus, say at 400°F. Drawing a horizontal line, we find that it intersects the liquidus at 40 per cent bismuth and 60 per cent cadmium. This composition of the liquid prevails all along the 400°F. level. We also know that because cadmium is the *excess constituent,* no bismuth has solidified. Therefore, we can set up the proportion:

$$\text{Cd:Bi::Cd:Bi}$$
$$x\text{:20::60::40}$$

where x = pounds cadmium still liquid = 30. Subtracting this from the original 80 lb. gives 50 lb. of cadmium that has solidified in the cooling of 100 lb. of an 80 cadmium–20 bismuth alloy to 400°F.

Review of Water-Salt System

And now let us turn back to review the water-salt system in the light of what we have learned in regard to the freezing of mixtures having eutectic points. A common and easily demonstrated example may clarify our knowledge of the less familiar, or at least less easily demonstrated, examples of metallic alloys.

Note that crystals of ice (frozen salt-free water) begin to form at the point *a* in the cooling of a 10 per cent solution of salt. More pure ice crystals continue to form as the mixture cools; and because of this separation of water, the brine gets continuously stronger. By the time it reaches the –7.6°F. temperature line, the brine consists of salt and water in the ratio of 23.5 to 76.5. It couldn't be anything else, because that is the proportional composition of the liquid at that temperature, the eutectic composition.

Let us suppose that our 10 per cent solution consists of 10 lb. of salt and 90 lb. of water. If we know that the solid, below –7.6°F., consists of ice crystals and eutectic crystals, how much of each constituent would be present? First, in regard to the amount of eutectic, we find that we have a mere 10 lb. of salt which can be used to make up this eutectic:

$$23.5:10::76.5:x$$
$$x = 32.55.$$

Thus the amount of eutectic is $10 + 32.55 = 42.55$ lb., and the amount of ice crystals is $90 - 32.55 = 57.45$ lb.

We learn from the diagram that below −7.6°F. any mixture of salt and water is solid and varies from 0 to 100 per cent eutectic, going from 0 to 23.5 per cent salt. We also learn that above 23.5 per cent salt a mixture of salt and water at that temperature consists of solid salt crystals and eutectic crystals. A thorough understanding of these facts is almost indispensable for mastering the iron-iron carbide diagram in the next chapter.

DEFINITIONS

The **liquidus** is the line on a constitution diagram where solidification begins on cooling. All temperatures above this line represent completely molten alloys.

The **solidus** is the line on a constitution diagram at which solidification is completed on cooling. Temperatures between the liquidus and the solidus represent a *mushy* stage, partly solid and partly molten. All points below the line represent completely solid alloys.

A **binary alloy** is an alloy containing two elements, apart from minor impurities.

The ASM "Metals Handbook" (1948) gives the following definitions:

Eutectic reaction. The isothermal reversible reaction of a liquid that forms two different solid phases (in a binary system) during cooling.

Eutectoid reaction. The isothermal reversible reaction of a solid that forms two new solid phases (in a binary system) during cooling.

Eutectic alloy. In any alloy system, the composition at which two descending liquidus curves in a binary system meet at a point. Thus such an alloy has a lower melting point than neighboring compositions. More than one eutectic composition may occur in a given alloy system.

REFERENCES

CLAPP, W. H., and D. S. CLARK, "Engineering Materials and Processes," International Textbook, Scranton, Pa., 1949.

DOAN, G. E., and E. M. MAHLA, "Principles of Physical Metallurgy," McGraw-Hill, New York, 1941.

"Metals Handbook," pp. 1146–1268, American Society for Metals, Cleveland, 1948.

SACHS, GEORGE, and KENT R. VAN HORN, "Practical Metallurgy," American Society for Metals, Cleveland, 1941.

TEICHERT, E. J., "Ferrous Metallurgy," 2d ed., vol. III, McGraw-Hill New York, 1944.

WILLIAMS, R. S., and V. O. HOMERBERG, "Principles of Metallography," 5th ed., McGraw-Hill, New York, 1948.

Chapter 5

THE IRON-IRON CARBIDE SYSTEM

A glance at the diagram (Fig. 31) of the iron-iron carbide system shows that it resembles Case IV except that its liquid-solid transformation resembles Case III. It will be noticed that there is a eutectic point *E* where the alloy of lowest melting point (4.3 per cent carbon) solidifies. It will be remembered that this is the carbon content of pig iron as it comes from the blast furnace, melting at 2065°F. It has been mentioned (see Air Furnace, Wrought Iron, etc.) that the purer the iron, or any metal, the higher its melting point. This fact may be noted on the diagram.

Interesting as this part of the diagram is, this text cannot cover it all. Attention will be directed mainly to that part between 0 and 2 per cent carbon; and for this reason, the diagram has been plotted with logarithmic abscissas to accentuate this area.

Notice that while the diagram is plotted horizontally in per cent carbon, this is for convenience only. It is really a diagram of iron-iron carbide relations; and iron carbide, the formula of which is Fe_3C, is the basis for study and calculations. This iron carbide is also known as *cementite*.

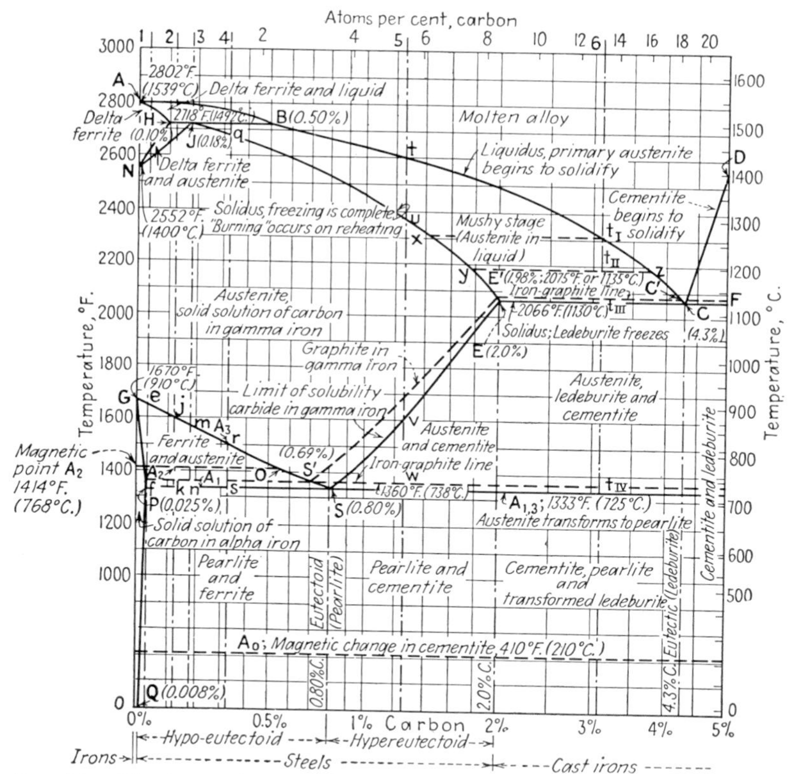

FIG. 31. The iron-carbon equilibrium diagram. (*By permission of American Society for Metals and R. S. Williams and V. O. Homerberg, "Principles of Metallography," 5th ed., p. 128, McGraw-Hill, 1948. Data include work of Kent R. Van Horn, John S. Marsh, F. N. Rhines, and J. H. Whitely.*)

Cementite

Cementite is the hardest of the several constituents occurring in iron and steel that will now be studied. It has been mentioned in some detail in connection with white cast iron. It is as brittle as glass. In fact, it will scratch glass, but it is not hard enough to scratch quartz. It is so hard to polish that when a specimen containing cementite is prepared for the microscope, the cementite particles stand out in relief.

Ferrite and Pearlite

Other constituents usually identified by the microscope can now be described in detail. *Pearlite* was also mentioned in connection with white iron (see page 43), where it was represented as consisting of alternate layers or laminations of cementite and ferrite. In most commercial ferrous products, this latter constituent contains alloys and impurities which, however, do not change its resemblance in appearance to pure iron. Its name, *ferrite,* indicates that it is nearly pure iron. Not only does this constituent exist in pearlite laminations; but as we shall see presently, it also is present as massive crystals, or grains.

Allotropy

Another constituent of steel is known as *austenite,* a definition of which is based on the fact that iron exists in two or more *allotropic* forms. This word *allotropic* is best understood by referring to the three different allotropic forms of carbon: soot, with a specific gravity of 1.88; graphite, with a specific gravity of 2.25; and diamond, with a specific gravity of 3.51. Thus it can be seen that it is possible for some of the elements to exist in more than one crystalline structure and density. Sulfur and phosphorus are two other instances of allotropy.

We shall consider two allotropic forms of iron—alpha (α) and gamma (γ). Transformations in iron and steel (refer to Case IV, Chap. 5) depend upon transitions from one allotropic form to another; and, generally speaking, gamma iron exists *above* the transformation temperature, while alpha iron exists *below* that temperature. These changes will be taken up in detail later and are indicated in Fig. 31.

Atoms of alpha iron arrange themselves in body-centered cubic

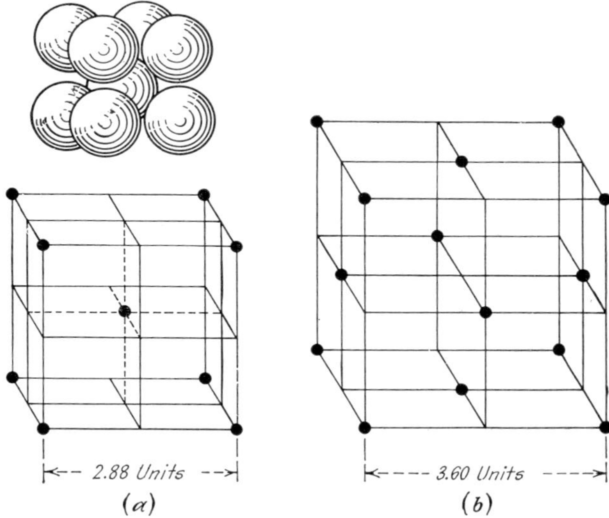

Fig. 32. Space-lattices illustrating *body-centered* arrangement of atoms at (*a*) and *face-centered* at (*b*). Notice that the atoms are packed much closer at (*b*). (*b*) may be described as consisting of two simple cubic lattices, corner atoms of one set being the face centers of the other set. (This can be visualized by building out into space.) Similarly, it can be seen that if (*a*) were extended out into space, the center atom shown above would act as corners for the next layer of cubes.

structure (see Fig. 32*a*). It can easily be seen that every atom has eight adjoining neighbors.[1] Atoms of gamma iron build themselves into a face-centered cubic structure (see Fig. 32*b*), a more closely packed structure because each atom has 12 adjoining neighbors.

This allotropic change alone would cause considerable atomic turbulence inside the metal, but when we also consider that gamma iron dissolves carbon up to 2 per cent, while alpha iron

[1] It must be clearly understood, of course, that atoms do not touch. They are vast distances apart compared with their size. However, their outer electron shells exert great repulsive force, and thus each atom can be considered as a comparatively large *sphere of influence*.

can hold only a minute amount[1] in solution, we should not be surprised at the temperature and volume changes that accompany a transition up or down through what is known as the critical range as indicated between the lines *GS* and *PS* and between the line *SE* and *PS* extended.

Austenite

Notice the area marked *austenite* on the diagram (see Fig. 31). This constituent has the same blank appearance under the microscope (see Fig. 33) no matter what its carbon content is (within the limits shown on the diagram). This indicates that it is a true solution.[2]

By the use of certain alloys which inhibit the natural change of austenite occurring at the 1333°F. line, this constituent can be brought down to ordinary temperatures without change and thus can be studied. In such a condition it can be hardened only by cold-working; it is so feebly magnetic that it cannot be picked up by a magnet. In short, it really resembles some soft metal such as copper more than it does the alpha iron or steel that we are accustomed to.

Eutectoid

Notice that the lines *GS*, *SE*, and the 1333°F. line resemble what we have learned to recognize as a eutectic reaction. This resemblance has caused it to be known as a *eutectoid reaction*, and its only difference lies in the fact that the changes occur within a solid instead of in the transition between solid and liquid. Likewise, a steel of composition indicated by

[1] Carbon is soluble in alpha iron to the extent of 0.008 per cent at ordinary temperatures and about four times that much at 1333°F.

[2] A solution may be defined as a homogeneous mixtures of two or more substances whose proportions may vary within fairly wide limits. It can be solid, liquid, or gaseous. A solution differs from a compound in that the latter is composed of substances in certain definite proportions.

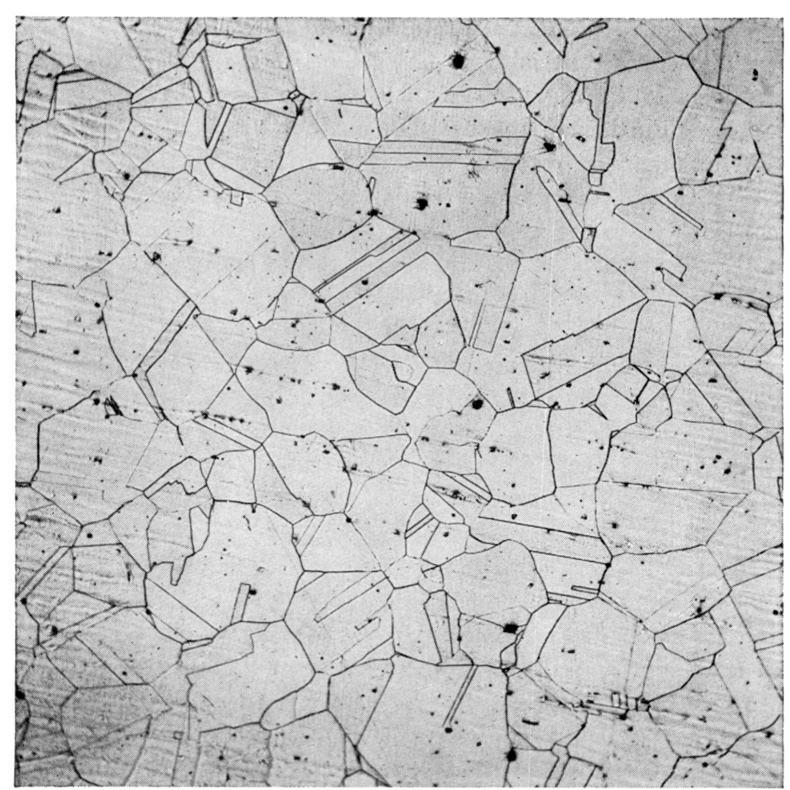

FIG. 33. Photomicrograph showing austenite in 18–8 stainless steel. Magnification 100X. (*Courtesy General Electric Company.*)

point S is known as *eutectoid steel.* See definition of eutectic alloy, page 67.

Transformations

The eutectoid composition is at 0.80 per cent carbon. Steel of this composition, on slow cooling, transforms 100 per cent from austenite to pearlite at the 1333°F. line. To the left of the 0.80 per cent point, there is an excess of ferrite over this eutectoid composition. Thus ferrite starts to separate out of solution (solid solution in this case) along the line *GS*, just as cadmium crystals start to form at the liquidus line in the cadmium-bismuth system. Further reference will be made to these transformations.

What happens when a steel of higher carbon content than the eutectoid composition cools down to the line *SE*? Any composition to the right of the eutectoid point has an excess of carbon over the eutectoid composition, just as there is an excess of bismuth to the right of the eutectic point in the cadmium-bismuth system. However, the constituent which separates out is not carbon.[1] Instead it is *iron carbide,* also known as *cementite,* or Fe_3C. The iron-carbon diagram is based on carbon content, but as long as we are speaking of steels (under 2 per cent carbon) and as long as there is no large amount of silicon present, the excess carbon separates out in the form of iron carbide.

That this Fe_3C is a distinct compound is attested by the fact that it has a definite composition, agreeing with atomic weights; and it has characteristics of its own differing greatly from those of its components taken separately. Also the two constituents are not separable, except by chemical reaction or by graphitization under the influence of certain alloys such as silicon.

As an example of the transformations mentioned above, let us slowly cool a piece of steel 0.80 per cent carbon content to see what happens. We might even begin with it in the molten condition. It would begin to solidify at 2680°F. go through a mushy stage, become entirely solid at 2530°F. and then gradually cool down to a trifle below the 1333°F. line. Here such a profound change would occur that the piece would

[1] This statement refers to steels in the ordinary use of the word. See definition of steel, p. 26. The separation of free carbon when the carbon and silicon are high enough, as in the case of malleable and gray cast iron, has already been described in Chap. 3.

brighten perceptibly and increase in temperature momentarily[1] while the atoms were taking on a new arrangement, passing from gamma to alpha iron. This does not account for all the disturbance either, for it happens that iron carbide is insoluble in

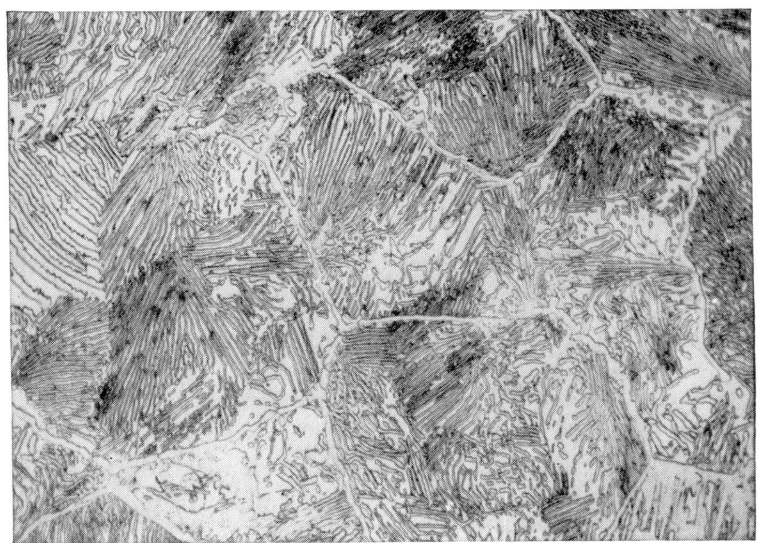

FIG. 34. Photomicrograph showing laminated structure of pearlite at magnification of 600 diameters on a 0.95 per cent carbon steel. As this carbon percentage is higher than 0.80, there is a small amount of cementite (the white areas). (*Courtesy General Electric Company.*)

alpha iron. The plain crystals of austenitic gamma iron change to a structure consisting of alternate plates of alpha iron and iron carbide; and to this structure the name *pearlite* has been given because of the way the ridges on a polished surface diffuse light, causing it to resemble pearl in appearance. (See microstructure, Fig. 34.)

[1] It is hoped that this can be demonstrated to the class by the experiment described on p. 235, in the Appendix.

Such laminated structures are incidentally, very characteristic of eutectic alloys (See Fig. 110).

Next, let us follow the slow cooling of a piece of steel of 0.20 per cent carbon. Steel of this composition would begin to solidify at about 2770°F. Solidification would be complete at 2700°F. and it would then be austenite down to where the 0.20 *C* line intersects the A_3 line—about 1525°F. At this point a change begins that resembles the formation of cadmium crystals at the liquidus of the cadmium-rich side of the cadmium-bismuth diagram. In this case crystals form in solid instead of a liquid solution.

What are these crystals? A glance at the diagram (see Fig. 31) shows that we are dealing with an area to the left of the eutectoid point, where a little consideration will convince us that there is an excess of iron over the eutectoid composition (0.80 per cent carbon). As is the case with all similar alloys, the excess constituent, in the form of ferrite crystals, will begin to separate out; and it will progressively separate out all the way down to the A_1 line. The remaining austenite (corresponding to the liquid portion of the cadmium-bismuth triangles) will become richer and richer in carbon, until at the A_1 line it discards all remaining ferrite in excess of the eutectoid ratio (0.80 per cent carbon) and then changes from austenite to pearlite at constant temperature. The microscope will show the patches of pearlite interspersed among the ferrite (see Fig. 35).

What has been described above takes place so unfailingly (provided the cooling is slow) that an experienced microscopist can estimate the carbon content very closely simply by comparing the relative areas of ferrite and pearlite. All white grains show nearly zero carbon. All patches of pearlite show 0.80 per cent carbon. This range, 0 to 0.80 carbon, is known as *hypoeutectoid,* meaning below, or beneath, the eutectoid percentage.

Let us now consider the hypereutectoid area, the region that

Fig. 35. Photomicrograph showing ferrite and pearlite grains in a 0.40 per cent carbon steel. Magnification of 100 diameters is not high enough to show the pearlite clearly. (White crystals are ferrite; darker ones are pearlite.) (*Courtesy General Electric Company.*)

has an excess of carbon over the eutectoid ratio, over in the *carbide-rich* part of the diagram, 0.80 to 2. Let us take the case of a 1.25 per cent carbon steel and follow its solidification and cooling. Solidification begins at 2600°F. and ends at 2350°F. Here we have a solid austenite solution of 1.25 per cent carbon. At about 1600°F. (the *Acm* line) a transformation begins. The austenite is beginning to be supersaturated with Fe_3C, cementite,

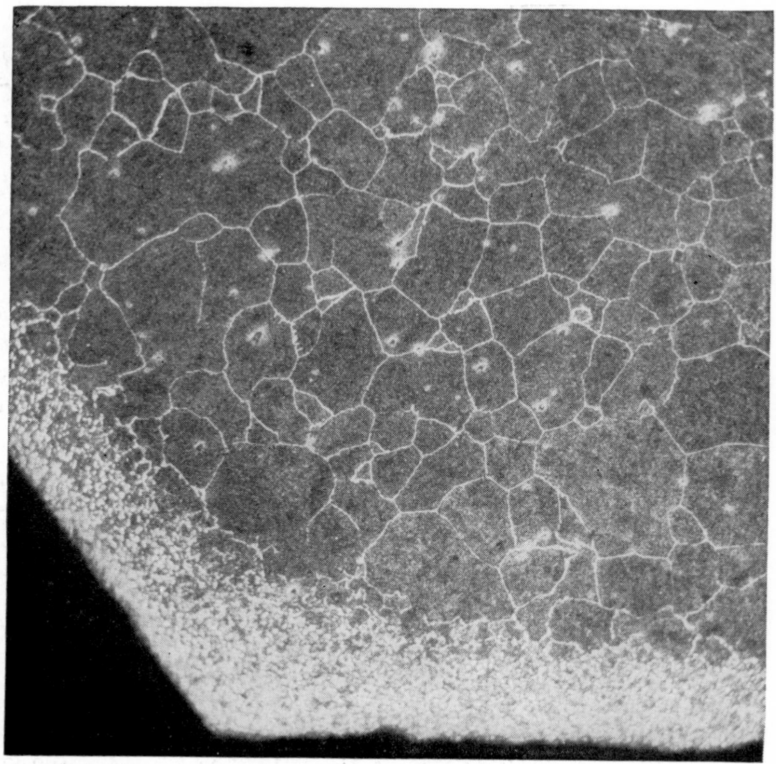

Fig. 36. Photomicrograph showing cementite in grain boundaries, hyper-eutectoid steel. Lower and left edges decarburized. (*Courtesy General Electric Company.*)

This constituent is expelled progressively all the way down to the 1333°F. line, when the remaining austenite transforms at that temperature into pearlite of the same structure and composition as the pearlite that was formed in the cooling of our previous case of 0.20 per cent or 0.80 per cent steels. In other words, the last remaining austenite in any steel cooling slowly down to 1333°F. is always of eutectoid composition.

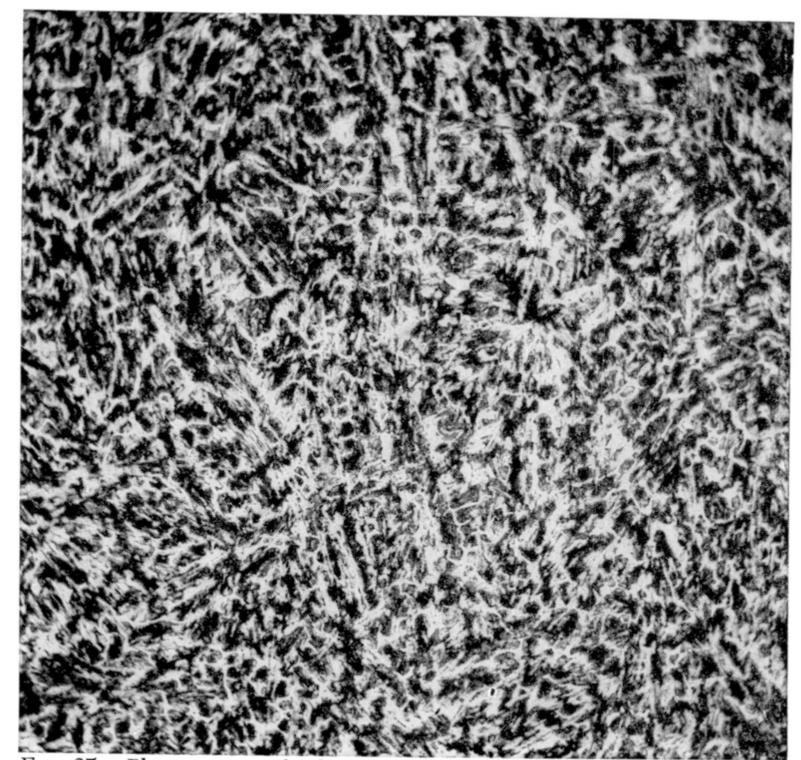

FIG. 37. Photomicrograph showing white cast iron, as cast. Low magnification, 60 diameters. Note the much greater proportion of cementite (white areas) than in Fig. 36, because of much higher percentage of carbon. Dark areas are pearlite. (*Courtesy General Electric Company.*)

If there is an excess of iron over that composition, it is expelled as ferrite crystals. If there is an excess of carbon, it is expelled as cementite. There is just one point of difference in this latter case; the Fe_3C is not expelled in the form of massive grains or patches but forms a network in the boundaries between the grains of changing austenite. This property of existing in grain boundaries (see Fig. 36, also Figs. 37 and 38) is what gave it

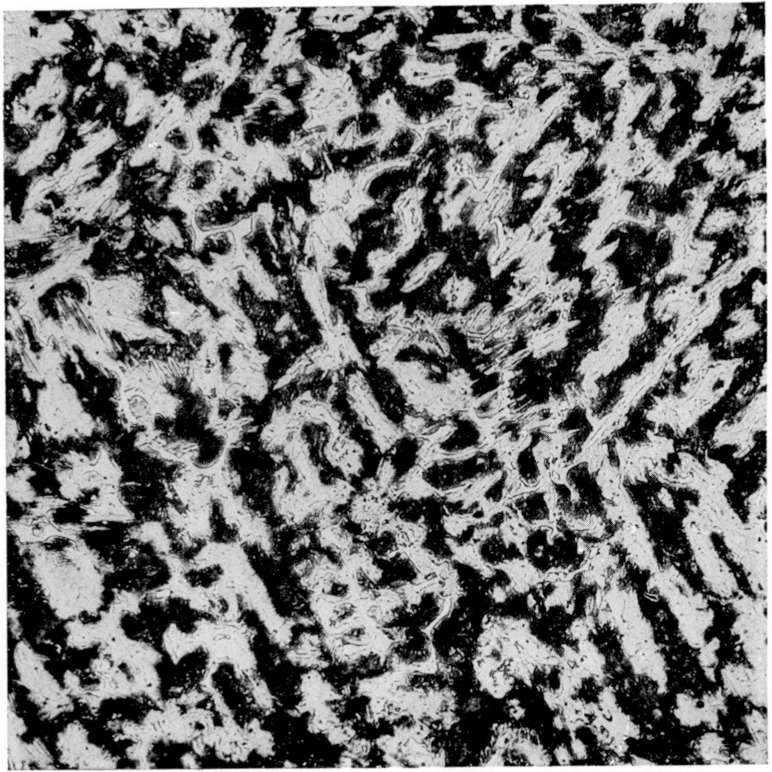

FIG. 38. White cast iron, as cast, showing pearlite with cementite in the grain boundaries. High magnification, 600 diameters. (*Courtesy General Electric Company.*)

its name of *cementite*—it seems to cement the pearlite grains together.

Transformation Range

Reference having been made to *A* lines, a little explanation is in order. It is obvious from the diagram that the *A* lines mark the beginning and end of transformations in a manner similar to liquidus and solidus lines; but, of course, those terms cannot be used where the transformations take place entirely in solid solutions.

The *A* originated from the French word *arret*, which was applied because the smooth cooling (or heating) curves were "arrested" or halted at these points. It must be remembered that the lines of the iron-iron carbon diagram were determined by points of change in cooling (and heating) curves just as in the case of any other alloy.

Changes from solid to liquid, or vice versa, usually occur at the same definite point. For instance, ice melts at 32°F. and water freezes at 32°F. Rarely is there a difference of more than a fraction of a degree between the two points. A change of structure in a solid, however, cannot take place so easily, and there is usually a delay; *i.e.*, the critical points in cooling are somewhat lower than those noted on heating (at least 20° in the case of steel).

The subscripts *r* and *c* therefore are used with the letter *A*, thus; *Ar* and *Ac*. The *r* comes from the French word *refroidissement* and designates A points on cooling, while *c* comes from the French word *chauffage* and designates A points on heating.

A little sketch is presented here to illustrate the *A* lines more clearly than on the large diagram. Unless special emphasis is required in differentiating between cooling and heating, the subscripts *r* and *c* are omitted. In this diagram only one *Ar* line is shown, and it is dotted.

A_1 represents the temperature at which gamma iron changes to alpha on cooling (or vice versa on heating). A_2 represents the line above which iron loses most of its magnetism. Notice that it merges with A_3, descends to the eutectoid point, and then continues along with both 1 and 3. A_3 represents the line where ferrite begins to separate from austenite. Notice that it descends to meet the A_1 line and coincides with A_1 to the right of the eutectoid point; this is in line with what has been said concerning hypereutectoid transformations, in that no ferrite crystals exist to the right of the eutectoid point.

There remains one more *A* line to describe. Its subscript, *cm*, stands for cementite, because it is along this line that cementite

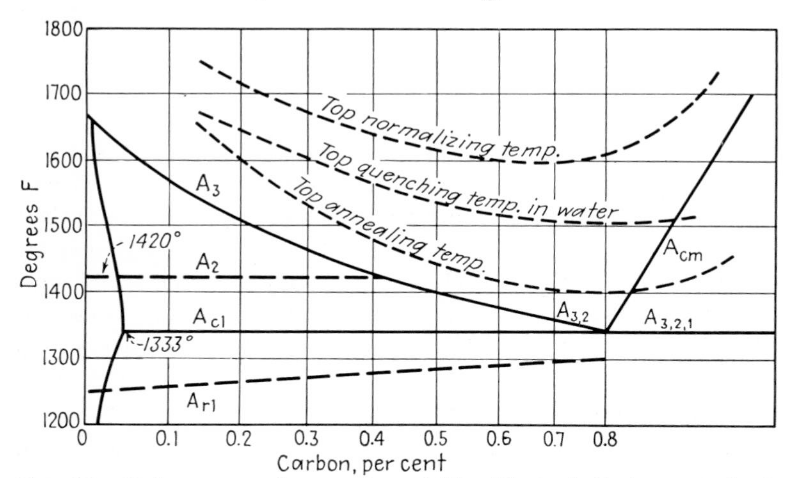

FIG. 39. Enlargement of a portion of Fig. 31, to indicate approximate heat-treating temperatures. Carbon per cent not plotted logarithmically.

begins to separate from austenite. The *c* of *cm* should not be confused with the *c* of *chauffage*.

Calculations

It is very convenient and enlightening to be able to determine the exact composition of a sample of steel. The formula Fe_3C means that $56 + 56 + 56$ parts of iron are combined with 12 parts by weight, of carbon. The exact percentage of carbon is $\frac{12 \times 100}{(3 \times 55.84) + 12} = 6.68$, but it is sufficiently accurate for our purposes to use the fraction $\frac{12}{(3 \times 56) + 12} = \frac{1}{15}$.

Thus, a 0.50 per cent carbon steel contains about $15 \times 0.50 = 7.5$ per cent Fe_3C; and, there being no free (massive) cementite, all this cementite is bound up in pearlite laminations, the ratio of cementite layers to ferrite layers being about 1 to 8.3. This means that a 0.50 per cent carbon steel, slowly cooled, has $0.50 \times 15 \times 8.3$ or 62.25 per cent pearlite. Thus, one-half of 1 per cent of carbon can enter into the structure of and strengthen 62.25 per cent or more of the iron it is combined with, more than doubling its strength, as mentioned in the opening paragraphs of Chap. 4.

A simpler method of calculating the amount of pearlite in any hypoeutectoid steel is to consider that a zero per cent carbon steel contains zero pearlite, while an 0.80 per cent carbon steel is composed of 100 per cent pearlite; and any intermediate percentage of carbon results in a proportionate percentage of pearlite. Calculating in this manner, a 0.50 per cent carbon steel would consist of 0.50/0.80 or 5/8 pearlite, or 62.5 per cent. The remaining 37.5 per cent would be ferrite.

Similarly, in calculating the amount of massive cementite (not interspersed as layers in pearlite), we know that there is zero per cent of this massive cementite in an 0.80 per cent carbon steel, while 6.67 per cent would represent 100 per cent (even though this high content is never reached) and intermediate amounts would contain massive cementite in percentages proportionate to their position in the range between 0.80 per cent and 6.67 per cent carbon. Thus, a 1.25 per cent carbon steel contains $\frac{1.25 - 0.80}{6.67} \times 100 = 6.75$ per cent of massive or excess cementite. The remaining 93.25 per cent would be pearlite.

DEFINITIONS

Cementite is a hard, brittle, crystalline compound, the composition of which is represented by the formula Fe_3C.

Austenite is defined in the "Metals Handbook"[1] as a solid solution in which gamma iron is the solvent." Its carbon content at maximum solubility (2066°F.) can vary from 0 to 2 per cent. This solubility decreases with decrease in temperature to 0.80 per cent at 1333°F.

Ferrite is defined in the "Metals Handbook" as "a solid solution in which alpha iron is the solvent," and which is characterized by a body-centered cubic-crystal structure." It is almost carbonless but may contain considerable silicon, nickel, phosphorus, and other alloys and impurities.

Pearlite is a decomposition product of austenite on slow cooling, containing about 0.80 per cent carbon and consisting of alternate laminations of ferrite and carbide.

Allotropy. The ability of a material to exist in more than one crystallographic structure is known as *polymorphism.* If the change is reservsible, it is *allotropy.*

A **hypereutectoid** steel is a steel containing more than the eutectoid percentage of carbon.

A **hypoeutectoid** steel is a steel containing less than the eutectoid percentage of carbon.

QUESTIONS

1. In what respects do the transformations at or between the *A* lines of the iron-carbon diagram resemble those at or between the solidus and liquidus of the cadmium-bismuth diagram? Name at least one point of dissimilarity.

2. Discuss the solubility of carbon in iron.

3. Suppose that a piece of plain-carbon steel was heated above the critical temperature (into the austenitic condition) and that as it cooled slowly, it showed one and only one transformation point: (*a*) What was its carbon content? (*b*) At what temperature did it transform? (*c*) What was its structure after transformation?

4. The instructor can assign a variety of problems in regard to calculating the percentages of ferrite, pearlite, and cementite for various carbon contents.

[1] "Metals Handbook," American Society for Metals, Cleveland, 1948.

5. Assuming pearlite to have a tensile strength of 125,000 p.s.i., and ferrite to withstand 50,000 p.s.i., calculate tensile strengths of slowly cooled plain-carbon steels of 0.10, 0.20, 0.30, 0.40, 0.50, 0.60, and 0.70 per cent carbon content.

REFERENCES

BRICK, R. M., and ARTHUR PHILLIPS, "Structure and Properties of Alloys," 1st ed., McGraw-Hill, New York, 1942.
"Metals Handbook," American Society for Metals, Cleveland, 1948.
TEICHERT, E. J., "Ferrous Metallurgy," 2nd ed., vol. III, McGraw-Hill, New York, 1944.
WILLIAMS, R. S., and V. O. HOMERBERG, "Principles of Metallography," 5th ed., McGraw-Hill, New York, 1948.

Chapter 6
HEAT-TREATMENT OF STEEL

The discussion, in the last chapter, of the constituents of steel clearly emphasized that slow cooling was necessary for their formation; and it was at least implied that, because time was necessary for the orderly and most natural arrangement of atoms, fast cooling would result in strained or unnatural arrangement.

Even on slow cooling, it was shown, there was a lag in transformation, as proved by the *Ar* points being somewhat lower than the *Ac* points. When it is considered that the changes in atomic arrangement in the transformation from face-centered gamma iron to body-centered alpha iron must take place in a *solid solution,* and that at the same time there occurs a great change in the solubility of carbon, it is not at all surprising to find structures with quite different qualities, according to the rate of cooling.

Hardness

Hardness is caused by this strained arrangement of atoms in steel or in any alloy. (This will be taken up in more detail under *grain size.*) Therefore, anything that increases such strains, such as faster cooling or greater percentage of carbon, increases hardness.

Hardness alone is seldom desired. If it were, white cast iron, Fe_3C, would be ideal, but its brittle quality makes it unfit for most uses. The process of heat-treatment is one involving both hardness and toughness, with both qualities controlled to meet the requirements of any case.

HARDENING

Nothing is accomplished in the way of hardening unless the steel is heated to a temperature that puts the carbon in solution. Heating to any point below the A_1 line causes no change in structure. Therefore, no matter how quickly the steel is cooled from 1300°F., no hardening will result. Quenching from a temperature between A_1 and A_3 will result in some degree of hardening because some transformation has taken place (see Fig. 40). Maximum hardness, however, results from a quench about 50° above A_3. Higher temperatures are not desired because of grain growth. Large grains cause brittleness. (Fineness of grain is usually desired in a steel, although it can be too fine. This will be discussed later.) The student may now begin to realize the importance of the iron-iron carbide diagram in the heat-treatment of steel.

Quenching Rates

In heating for quenching, the size of the piece is a factor to be considered. Heating too quickly may result in some part being overheated or distorted and perhaps cracked. When it has reached the proper temperature, it must be held there long enough to have a uniform temperature throughout. One hour per inch of penetration is a good rule.

Different rates of cooling have definite effects on structure and properties. Let us take a steel of 0.45 per cent carbon as an example. If it is heated above the A_3 line, as mentioned above, and then cooled slowly, say in hot ashes, the structure will be coarse, lamellar pearlite and some ferrite crystals. It

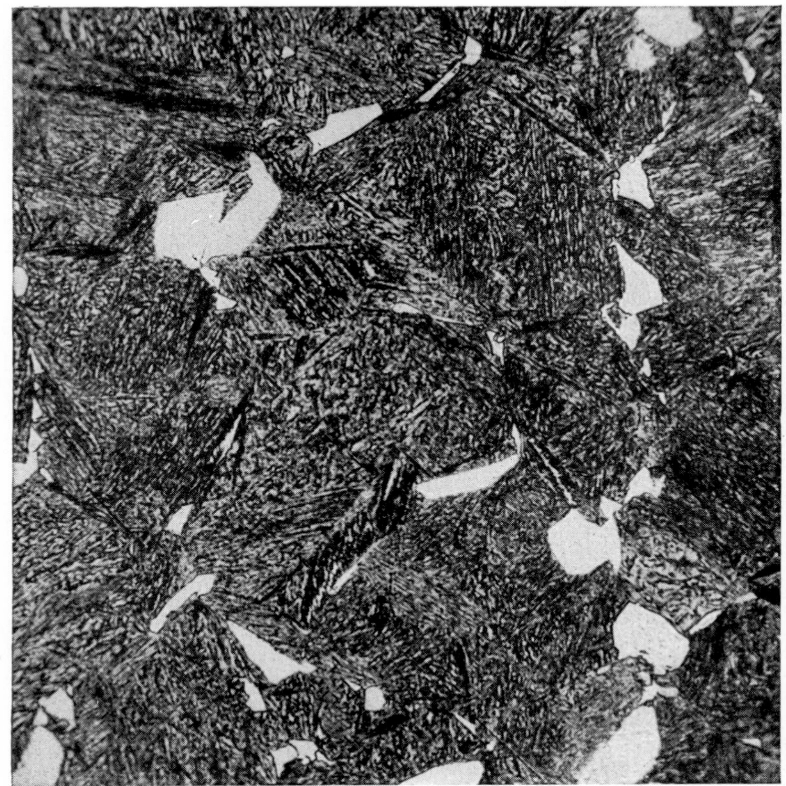

FIG. 40. Photomicrograph taken at 750 diameter, showing result of a quench from between A_1 and A_3 lines. Lack of sufficient heat resulted in incomplete transformation. White particles are ferrite, which weakens the structure. Dark area is martensite. (*Courtesy General Electric Company.*)

will be in a soft condition, having a Brinell hardness number[1] of about 150, a tensile strength of about 90,000 p.s.i., and an elongation of about 25 per cent. Next, let us consider what would happen to the same steel if instead of being slowly cooled it were given a mild quench, say in a blast of air. The cooling

[1] Explained in the chapter on Testing.

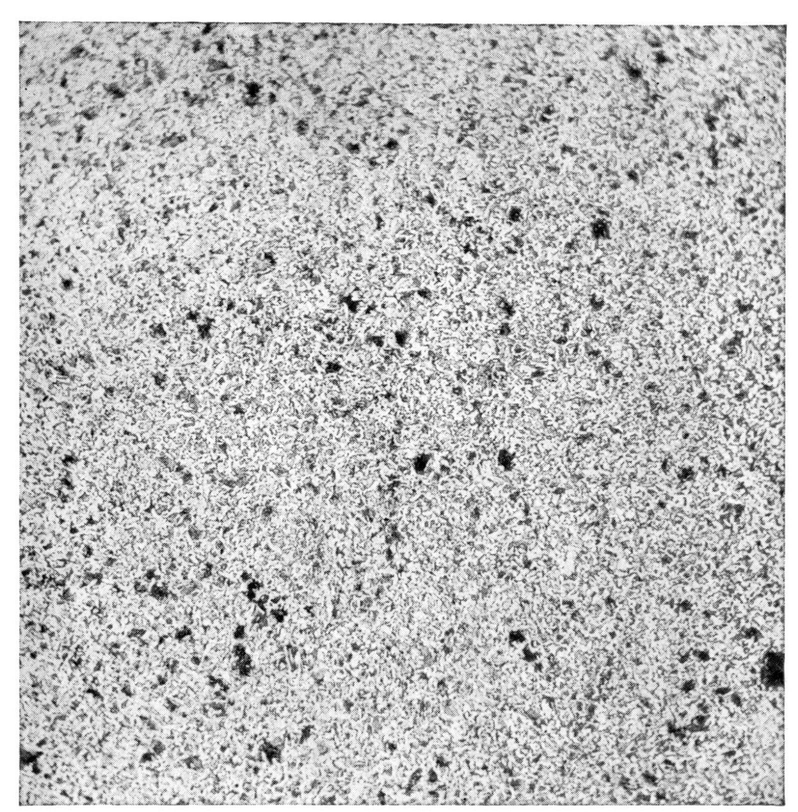

FIG. 41. Photomicrograph showing 0.30 per cent carbon, 1.18 per cent manganese, and 0.15 per cent vanadium steel, air-cooled from 1650°F. Magnification 100X. (*Courtesy General Electric Company.*)

is too fast for the coarse pearlite laminations to form. Instead a finer and more dispersed pearlite results (see Fig. 41). As a result of the more sudden cooling, the austenitic transformation point has been lowered to, say 1200°F., and the dispersion of the carbide particles has resulted in a decreased amount of free ferrite, resulting in a higher Brinell hardness of 200; a slightly

FIG. 42. Martensite at 500 diameters. (*Courtesy General Electric Company.*)

higher tensile strength of 96,000 p.s.i.; and, not being as ductile, a lower per cent elongation of 22 per cent.

Next, if a considerably faster quenching medium were used, it would be quite possible to obtain pearlite so extremely fine that it appears as a black mass under usual magnification. Here again there are increases in Brinell hardness and tensile strength and a decrease in elongation.

Finally, if the steel of our experiment were given a very fast quench, say in cold water, a constituent would result that possesses the maximum hardness, Bhn 400 to 700; tensile strength, 180,000 p.s.i.; and very low ductility, as shown by an elongation of about 10 per cent. This constituent is very easily identified. A good description would be *needlelike* (see photo-micrograph, Fig. 42). The term *acicular* is often used to describe this structure. It may appear either light- or dark-colored.

Brinell numbers may not mean anything definite to the beginner in metallurgy. Let us say that a file is a good example of martensite; and a plain-carbon steel heat-treated so that a file will barely mark it might be the dark mass described above.

What has just been said in regard to obtaining properties and structures by regulating the rates of cooling is all right in theory but, except in the case of very small pieces, is a difficult way to obtain them in practice. One reason for this is that a piece of appreciable size, say 2 in. in cross section, will not cool uniformly. When quenched in water, the outer surface will chill to form martensite, but the heat cannot travel outward fast enough to accomplish a quick cooling of the interior. Furthermore, with the center still warm, there is a continual heat transfer to the surface which will modify or even destroy the martensite, that was first formed. Another reason is that temperatures of quenching mediums are difficult to maintain.

TEMPERING

A more certain method of obtaining any desired strength and ductility is to quench the steel from above the critical fast enough for maximum hardness, which is martensite (often a water quench in the case of carbon steels and quite likely oil in the case of alloy steels), and then reheat to relieve stresses or to modify the martensite according to the quality desired.

FIG. 43. Photomicrograph showing sorbite obtained by tempering martensite, 500 diameters. At lower magnification it appears as a granular mass. (*Courtesy General Electric Company.*)

Such a reheating, to some point below the A_1 line, is called *tempering.* This term is familiar to most people. Perhaps they have been cautioned to be careful in sharpening tools, so as not to *draw the temper* by getting them too hot.

Reheating to 400°F. or lower does very little to modify martensite, but it accomplishes a great deal in relieving stresses caused by unequal cooling in the quench, stresses that are often severe enough to crack the steel.

Reheating above 400°F. results in a gradual transformation of the acicular martensite to a more granular form. At first the carbide is dispersed in particles of such small size as to be discernible only through microscopes of the highest power. At one time this stage was known as *troostite,* and is still mentioned in the "Metals Handbook" (1948) as "tempered martensite that etches rapidly, usually appears dark and is not resolvable by the microscope." This stage is now usually spoken of as *tempered martensite;* if it is desirable to specify the degree of temper, that can best be done by specifying the temperature to which the tempering should be carried.

When the tempering temperature has reached 750°F., the carbide particles have agglomerated enough to be distinguished as granules which grow larger and more definite as the temperature increases. To such a structure the name *sorbite* was given, and this term is still in use to a considerable extent.

Physical Properties Charts

As was stated at the beginning of this heading, tempering after a martensitic quench provides a quite certain method of obtaining the quality desired; and as an aid in deciding what steel to use for any certain job, several steel companies have prepared "Physical Properties Charts." These charts are very valuable in making decisions not only as to what steel to use but also how it should be heat-treated. They are to be used as guidance and are also valuable as an adjunct to the study of metallurgy, but they should not be considered as absolute values in every case. The Bethlehem Steel Company Handbook No. 268, "Modern Steels and their Properties" (1949), from which the several charts in this book were reproduced, states: "The data shown represent average values for the compositions and are not to be interpreted as the maximum or minimum obtainable."

Comparisons of the charts show the effect of variation in carbon content, tensile strengths being higher with increase in

AISI-C 1040, Fine Grain

(*Water Quenched*)

PROPERTIES CHART

(Average Values)

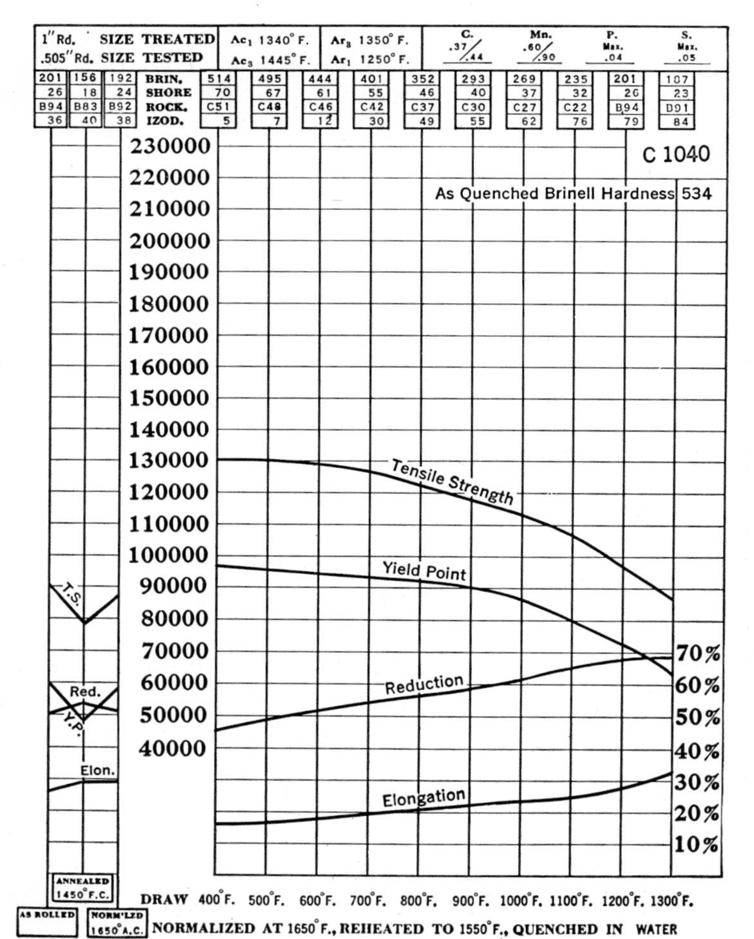

FIG. 44. Properties Chart. (*Reprinted from "Modern Steels and Their Properties," Handbook 268, Bethlehem Steel Company, 1949.*)

AISI-C 1040, Fine Grain

(Oil Quenched)

PROPERTIES CHART

(Average Values)

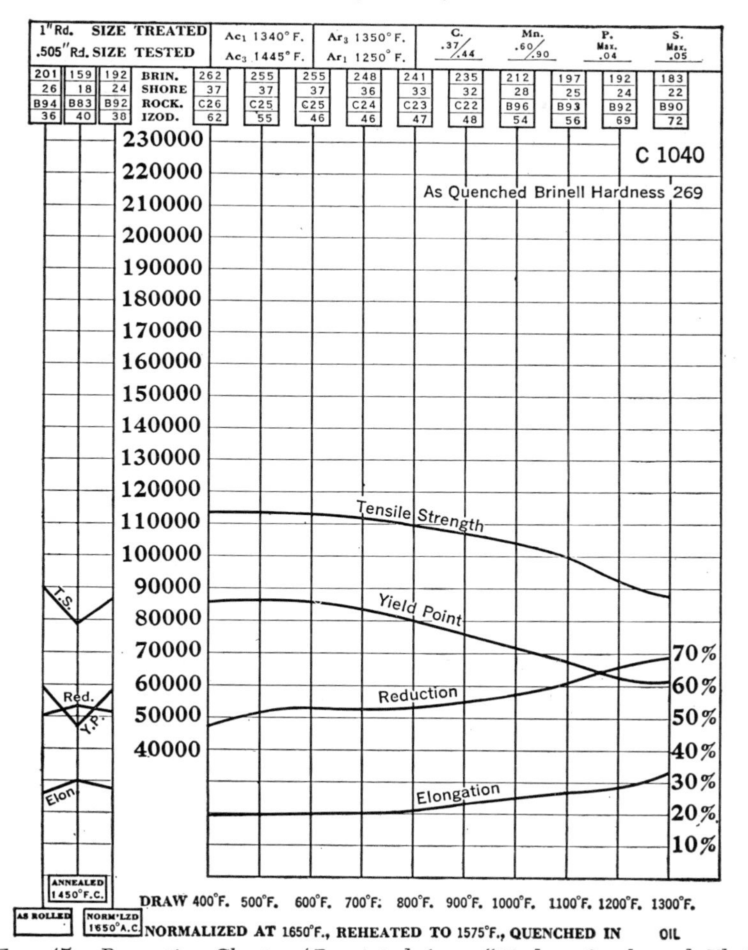

FIG. 45. Properties Chart. (*Reprinted from "Modern Steels and Their Properties," Handbook 268, Bethlehem Steel Company, 1949.*)

AISI-C 1080, Fine Grain

(*Oil Quenched*)

PROPERTIES CHART

(Average Values)

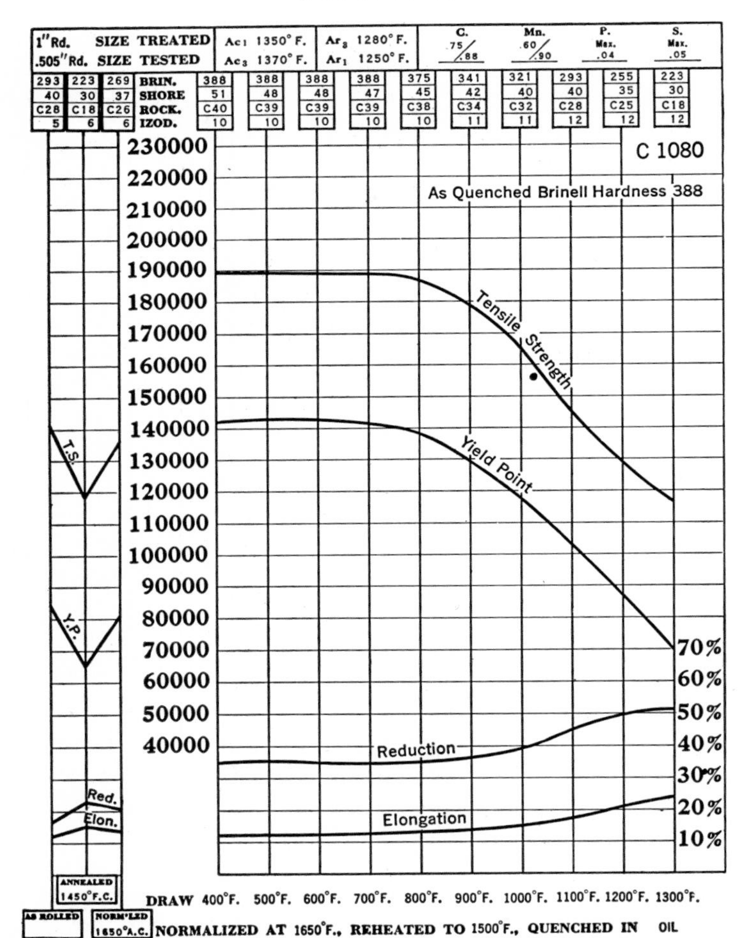

FIG. 46. Properties Chart. (*Reprinted from "Modern Steels and Their Properties," Handbook 268, Bethlehem Steel Company, 1949.*)

carbon content. Also the comparative effects of water and oil quench can be easily seen. In addition, these charts contain information which will be enlarged upon in later chapters. Charts with similar carbon content and heat-treat are included in the chapter on Alloy Steels, to show the effect of the various alloys.

It might be well to call attention to a few items of information that can easily be overlooked. For instance, the columns at the extreme left indicate values for annealed and for normalized steels. The letters *F. C.* mean furnace-cooled, while *A.C.* means air-cooled. Also note the chemical composition and the A_c and A_r temperatures at the top of the charts.

Tempering by Color

In all production work, the desired temperature is maintained by molten salts or lead; but it might be interesting, before closing the chapter, to list the *temper colors* (colors taken by iron oxide) by which yesterday's blacksmiths used to work and which are still used extensively. They are the natural colors that a bright iron surface takes when exposed to air at the temperature indicated:

400°F. Faint straw
440°F. Straw
475°F. Deep straw
520°F. Bronze
540°F. Peacock
590°F. Full blue
640°F. Light blue

A classic example of quench and of temper by color was the old blacksmith's method of heat-treating a cold chisel. He shaped it and ground the edge and then heated it to a light cherry. He then plunged it in water but withdrew it while still hot, sandpapered the edge so that the temper colors would show plainly, and then waited for the residual heat in the body of the chisel to travel to the edge and heat it. The edge would first show a straw color, then a bronze, etc. When it reached full blue, he would plunge the whole chisel in water. This stabilized the structure at the edge at the greatest hardness it could have and still be tough enough to withstand shock. As for the body of the chisel—because of its mass, it was not fully quenched in the first quenching; and even if the surface was quenched from above the critical, the residual heat in the interior would temper it. The net result was to have a good hard-cutting edge and a shank of steel soft enough not to shatter when struck by the hammer.

REFERENCES

Bullens, H. K., and Metallurgical Staff of Batelle Memorial Institute, "Steel and Its Heat Treatment," Wiley, New York, 1948.

Grossman, M. A., "Principles of Heat Treatment," American Society for Metals, Cleveland, 1940.

"Metals Handbook," American Society for Metals, Cleveland, 1948.

Sisco, Frank T., "Modern Metallurgy for Engineers," Pitman, New York, 1948.

Chapter 7
HEAT-TREATMENT OF STEEL (Continued)

SPHEROIDIZING

In the previous chapter, the tempering of martensite was described from its beginning, with the steel in a hard and brittle condition and with carbide particles in such a fine dispersion as to be indistinguishable, on through a gradual decrease in hardness and increase in ductility with increasing temperature. This was accompanied by a coalescing of the carbide particles until they became granules easily seen with a microscope.

If now this heating is continued until the steel is near a temperature of 1333°F., the critical temperature, a new force makes itself felt. There is enough mobility at this temperature so that the particles can contract their surfaces and become spherical. The same force, surface tension, causes falling raindrops to become spherical.[1]

The terms *tempering* and *drawing* are not applicable to this condition, although they merge into it. Considerable time is required for *spheroidization.* Not only is tempered martensite capable of spheroidization, but the process can be applied to pearlitic steels and steels with massive cementite (hypereutectoid). Coarse pearlite presents quite a difficulty because of the

[1] The ratio of the surface of a cube to that of a sphere of equal volume is about 5 to 4.

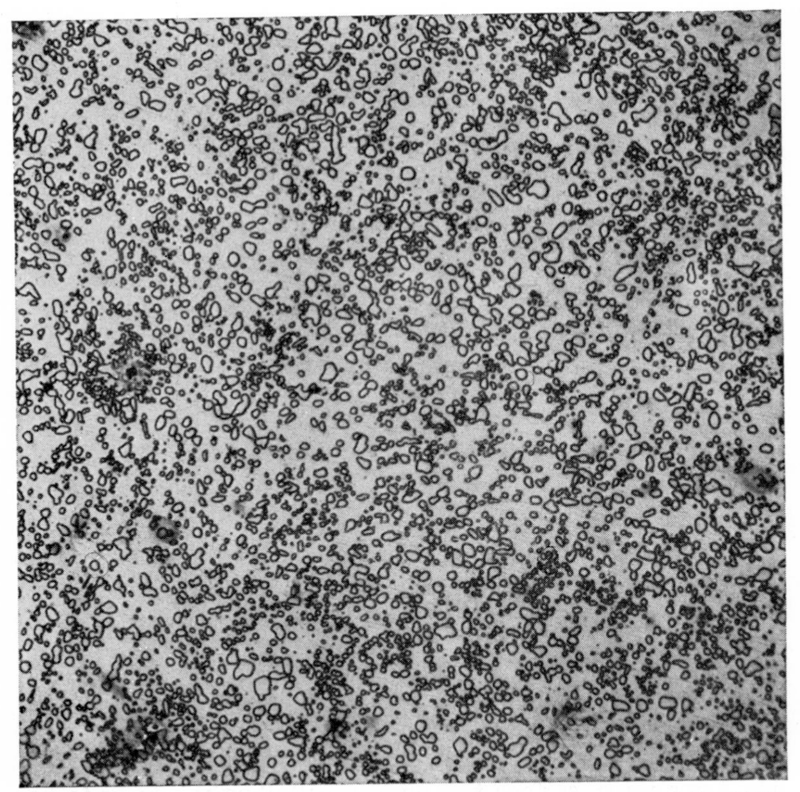

Fig. 47. 1.30 per cent carbon steel, spheroidized. Magnified 300X. (*Courtesy General Electric Company.*)

size of the carbide laminations. *Normalizing* (see page 102) before spheroidizing is often advisable in such cases. Spheroidized steel is often softer than annealed steel if the pearlite is coarse. It is much more easily machined, as the sketch (see Fig. 48) attempts to show. The tool meets less resistance sliding between carbide particles than in cutting transversely across pearlite plates. In obtaining the maximum ductility and softness by spheroidizing, there is some loss of strength and elasticity, but the loss is not great and can usually be taken care of in design.

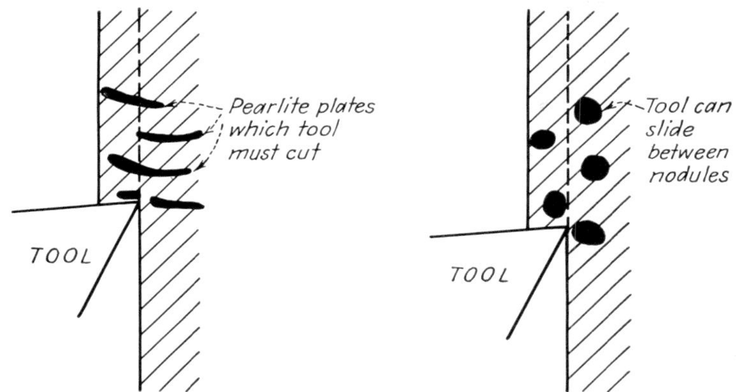

FIG. 48. Sketch illustrating greater ease of cutting spheroidized steel as compared with pearlite.

ANNEALING

Annealing is a rather general term used to convey the idea of softening. It is usually applied to the process of softening hard steel so that it may be machined. This is accomplished by heating material to about 50°F. above the A_3 line in hypoeutectoid steels or the A_1 line in hypereutectoid and then cooling slowly either in the furnace or in some heat-insulating material like dry lime.

Of course, the steel should remain at temperature long enough for transformation into the austenitic condition in the case of eutectoid and hypoeutectoid, but it should not he heated as high as the *coarsening temperature* of that steel. This temperature varies considerably with different steels. Aluminum-killed steels have a fine-grained structure that is stable up to higher temperatures than aluminum-free steels, while the grain size of the latter increases rather uniformly with increase in temperature, and they should be heated no higher than necessary. As to hypereutectoid steels, they should not be heated much above their A_1 temperature.

The *Acm* line ascends so steeply that complete conversion to austenite would result in entirely too large a grain structure. Besides, in the usual run of steels, the excess cementite does not run over 5 per cent, and the pearlite representing the main part of the steel is already in solution above the A_1 line. Grain growth is discussed more fully in Chap. 9.

As was stated, the word *annealing* is in use as a rather general term and includes such operations as heating up to below the critical in the manufacture of wire, when drawing has made the wire too brittle for further reduction. Several so-called *anneals* are sometimes necessary before the wire is down to size. (See Fig. 81, page 149 for a description of wire drawing.) Such anneals are not *full anneals,* but they do come under the head of *softening*. *Stress relief* by comparatively low temperatures, especially after cold-working, is also spoken of as an *annealing* process.

NORMALIZING

In cases where abnormally large grain structure is present, such as in steel castings, or where the structure produced by rolling or forging is undesirable, it is often necessary to heat the steel to a higher temperature than that for quenching or annealing before the carbides are in complete solution and the undesirable structure is eliminated.

Steel castings, especially large ones, because of their slow cooling, often have not only a large grain structure, but there may be quite a segregation of constituents (see Fig. 61). The treatment consists of heating to a temperature indicated by the dotted line, shown as normalizing temperature on the diagram, Fig. 39, long enough to get the constituents in solution, and then cooling through the critical range in still air at ordinary temperature. This treatment should result in a homogeneous fine pearlite structure without much of either ferrite or massive cementite, and this should satisfactorily respond to any further heat-treatment that may be necessary. Notice that most of the Physical Properties Charts reproduced in this book were normalized before heat-treatment.

TIME, TEMPERATURE, TRANSFORMATION DIAGRAMS

In pages 87 and 91, we mentioned the influence of the rate of cooling on the transformation of austenite and the various structures formed. The investigations of Bain and Davenport have helped greatly to put the heat-treatment of steel on a sound, scientific basis. In the now well-known S *curves*, or T-T-T diagrams, temperature is plotted vertically, against time on a logarithmic scale in the horizontal direction (see Fig. 49). The S to the left marks the beginning of transformation of the austenite, while the one on the right marks the end of austenitic transformation. Note that the 1333°F. line, with the designation "stable austenite" above it, ties in with the iron-iron carbide diagram. The significance of the area below the 1333°F. line, also marked as "stable austenite," will be explained in this chapter. Figure 49 was constructed for 0.89 per cent plain-carbon steel. The shape of the S does not vary greatly for other plain-carbon steels, but when it comes to alloy steels, there is a great deal of difference (see Fig. 89).

Studies leading to the construction of S curves[1] were con-

[1]Adapted from R. S. Williams and V. O. Homerberg, "Principles of Metallography," 5th ed., McGraw-Hill, 1948.

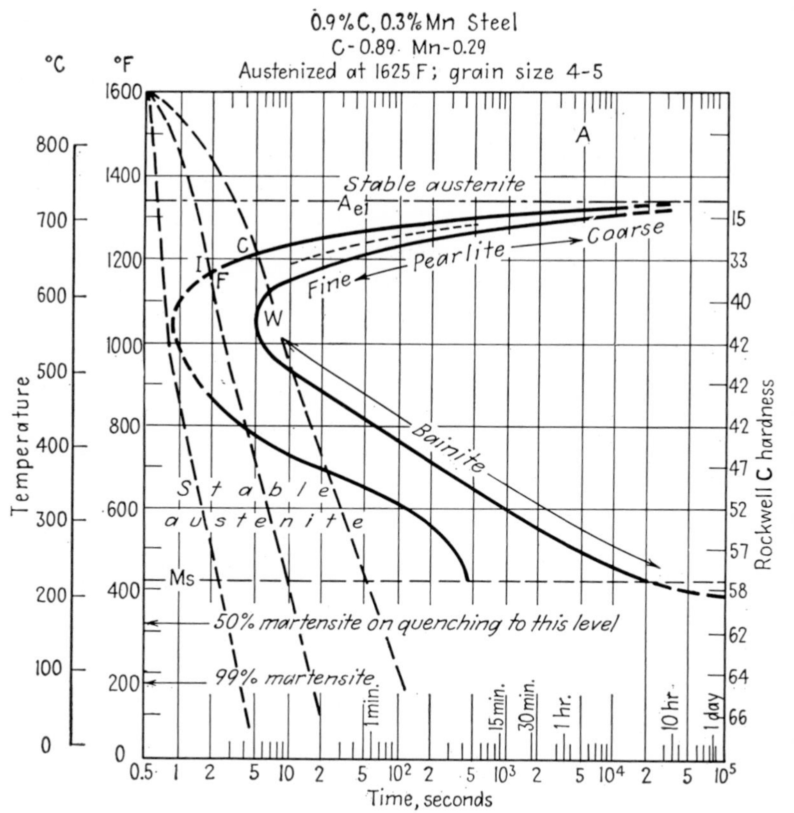

FIG. 49. S curves for 0.89 plain-carbon steel. (*Reprinted by permission from E. S. Davenport, from "Isothermal Transformation Diagrams in Steel," "Metals Handbook," p.* 608, 1948.)

ducted by rapidly cooling steel specimens from the austenitic state at temperatures below that at which the steel transforms on slow cooling. The rate of transformation and the products of transformations at these temperatures were noted. The procedure is to quench small specimens from a temperature at which austenite is normally stable into molten lead or a low-melting-point alloy kept at the lower temperature under consideration. These specimens are then removed from the bath at definite time intervals and quenched in water. By this procedure any untransformed austenite is converted into martensite. A microscopic examination discloses the extent of the austenitic transformation after quenching in water. Finally, the time required to complete the transformation is determined for each one of the lower temperatures.

In studying the diagram, first let us follow through what happens in a very fast water quench. This is indicated by the dash line at the extreme left. If the quench is fast enough to descend to 900°F. in 1 sec., in order to get by the "knee" of the S, then transformation will not begin until the *Ms* line (about 400°F) and will be complete by the time ordinary temperatures are reached. The result will be martensite. Note that the whole operation is over in a matter of seconds. No wonder that martensite has a stressed or unstable structure.

Second, let us follow through what would happen in a quench that is much slower, say ten times as slow, but still considered as a relatively fast quench. This is represented by the dash line *I*. Notice that in a few seconds the temperature was lowered to about 1180°F. In this case the cooling curve (the dash line) did not get by the knee of the S. It entered the zone of transformation at *F* and partly changed to the structure forming at that temperature—to very fine pearlite. It was cooling too fast for complete transformation at that point. The dash line did not reach the right-hand S. As the remainder cooled down to the *Ms* line, it was still austenite. This austenite transformed to martensite below the *Ms* line, and the result at ordinary temperature was a mixture of martensite and fine pearlite.

Third, let us follow through a cooling rate as represented by the dash line *C*. This cooling rate was too slow for the formation of any martensite or, indeed, of any of the finest pearlite.

The austenite began to transform at *C*, and this transformation was complete at the point *W*, the resulting structure being a medium-fine pearlite (see Fig. 41).

Finally, let us see where very slow cooling, such as was specified in Chap. 6, would fit into the S diagram. A line drawn to indicate a cooling rate of about a day would go through the transformation at about 1290°F., and the result would be ordinary coarse pearlite. It should be remembered that the diagram shown by Fig. 49 is for a carbon content near the eutectoid point. For a hypoeutectoid steel, a third line, partly sketched in, could be included to show where the formation of the ferrite ends and the formation of pearlite begins.

The knowledge gained and set forth in the T-T-T diagrams has not only clarified the whole matter of quenching and hardening, but has also led to at least two new methods.

Austempering and Martempering

In *austempering*,[1] the steel, as is usual in heat-treating, is heated to a temperature at which it is in the austenitic condition (austenitized) and quenched rapidly past the knee of the S curve; but instead of continuous cooling to ordinary temperature, it is quenched into a medium which is held at some intermediate temperature, say between 500°F. and 800°F., and held at that temperature for a period of time as indicated by the S curve for *complete transformation*. Note that this is a constant-temperature transformation. It is often termed *isothermal, iso* meaning "same."

Austempering is usually carried on by quenching in *salt baths*. These are molten salts such as mixtures of sodium carbonate, barium chloride, and others, held at correct temperatures by

[1] The principles of austempering are set forth in U.S. Patent 1,924,099, granted to Bain and Davenport in 1933; also in subsequent patents held by U.S. Steel Corporation.

thermostatic control. If the steel were quenched at 600°F., about 100 sec. would elapse before transformation would begin, and 1,000 sec. before it was completed. The resulting structure would be an acicular structure resembling martensite as viewed under the microscope. It has been termed *Bainite* in honor of Dr. E. C. Bain. Such a structure is hard, 50 to 55 Rockwell C, or a Bhn of 500 to 540, but superior in ductility and toughness to that obtained by the conventional quench and temper to the same hardness.

Size has been a factor in austempering because in large work there is too much difference in temperature between surface and center to obtain the full benefit of isothermal quenching. However, according to the U.S. Steel booklet, "U.S.S. Carilloy Steels," a modification of the above process has been developed which extends its use to larger sizes and heavier sections:[1] "The steel is quenched rapidly from above the critical, past the knee of the S curve, to some temperature below the *Ms* line, where small amounts of austenite transform to martensite, but is then transferred immediately to a bath or furnace held at a temperature desired for a Bainite microstructure."[2]

Martempering

Martempering[3] consists of a quench fast enough to get by the knee of the S curve, down to some temperature above the *Ms* line, where it is held a short time for the purpose of equalizing surface and center temperatures and then slowly cooled in air to produce martensite. The purpose is to avoid the distortion and cracking that a fast quench all the way down to ordinary temperature might entail. An immediate tempering operation is advisable.

[1] "U.S.S. Carilloy Steels," p. 106, Carnegie-Illinois Steel Corporation.

[2] U.S. Patent 2,258,566.

[3] B. F. Shepard, Martempering, *Iron Age*, Jan. 28, 1943, and Feb. 4, 1943.

Figures 50*a*, 50*b*, and 51 illustrate graphically the three methods of heat-treatment.

Figures 52 and 53 are included to show the martempering operation with respect to the S curves.

Figure 52 shows that in the ordinary fast quench, fast in order to get by the knee of the S curve without change,

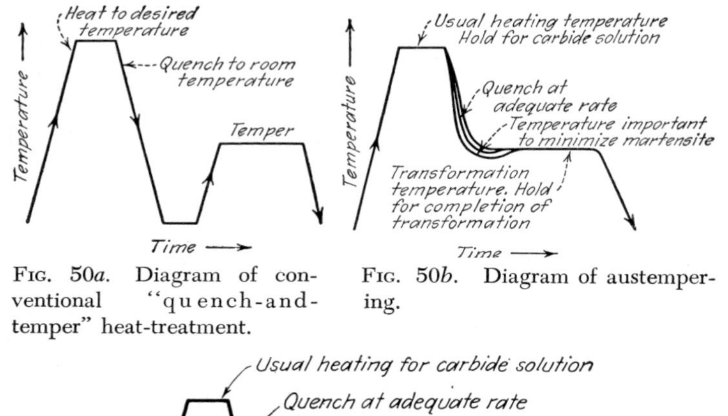

FIG. 50*a*. Diagram of conventional "quench-and-temper" heat-treatment.

FIG. 50*b*. Diagram of austempering.

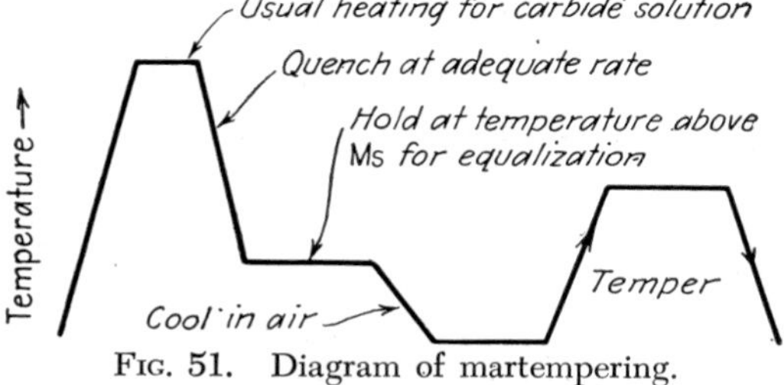

FIG. 51. Diagram of martempering.

quite a temperature difference between surfaces and center develops, a condition very conducive to distortion and cracking if the fast quench is continued to completion. Martempering avoids this danger by slowing up the latter part of the quench, thus allowing the temperature of the surface and interior to equalize. A similar objective is often practiced by heat-treaters when they quench steel in brine for one second and then finish the quench in oil.

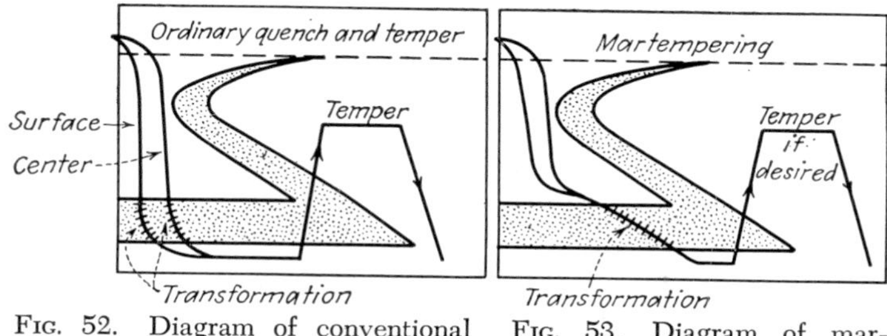

FIG. 52. Diagram of conventional quench and temper, with respect to S curve.

FIG. 53. Diagram of martempering, with respect to S curve.

Additional S curves are included in the chapter on Alloy Steels to show the effect of alloys on the shape of these curves.

SURFACE-HARDENING TREATMENTS

It is very often desirable to have a steel part very hard and resistant to wear on its outer surface and at the same time tough and comparatively soft in the interior. Often it is machined to shape in the soft condition; and after machining, the outer surface is hardened by one of the following processes: *carburizing, nitriding, cyaniding,* and also *induction* and *flame hardening.* These processes all come under the general term *casehardening.*

Carburizing

In *carburizing,* the pieces are packed in a box with carbonaceous material, usually a mixture of charcoal, barium, calcium, and sodium carbonates; the box is sealed and heated above the Ac_3 temperature, usually to about 1700°F. or 1800°F. At this temperature the austenite absorbs the carbon, the penetration being from 1/32 to 3/32 in., thus resulting in an outer "case" of high-carbon steel.

The steel does not absorb carbon from the solid state; rather the mechanism of absorption is as follows: A little oxygen is

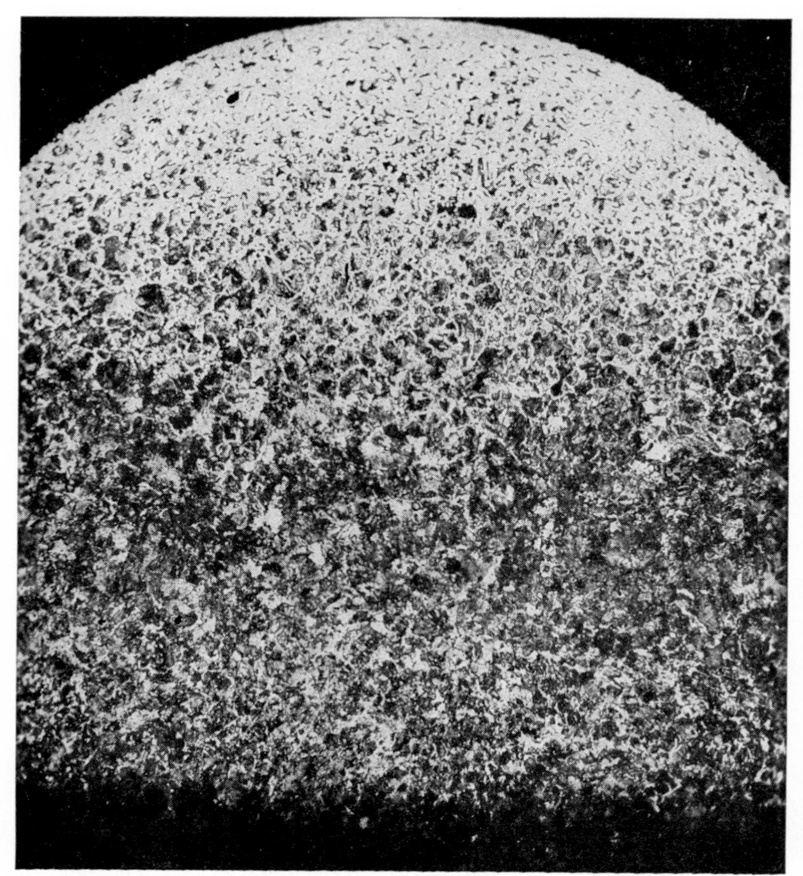

FIG. 54. Photomicrograph of carburized case, showing gradations from low to high carbon, going from top to lower edge. Beginning at the top, note, first, mostly ferrite grains (low carbon); second, as carbon content increases, the proportion of pearlite increases up to a point where it is 100 per cent. From there on downward to lower edge (outer edge of the piece) the cementite lacework appears.

entrapped in the sealed box. This oxygen burns to CO in an excess of carbon. It is from this gas that the austenite takes up its carbon content. After robbing the CO of its C, the freed oxygen unites with some more carbon and carries it over to the iron. Oxygen is thus the carrier. From this description, it will be seen that CO gas can be an effective carburizer. Illuminating gas is also used.

Carburizing results in a deeper case than the other methods. The quenching from 1700° may leave both core and case coarse-grained. Two reheatings and quenchings may be needed for refining the grain in both case and core, one from just above the A_3 of the low-carbon core and the other from just above the A_3 of the high-carbon case. Many fine-grained steels are quenched directly from the carburizing operation. It is interesting to note the gradation in carbon content in the *case* of a carburized piece of steel as seen by the microscope (see Fig. 54). The carbon content is greatest at the surface, and it tapers off gradually toward the interior.

Nitriding

Nitrogen, like carbon, also hardens iron, and so a process of casehardening known as *nitriding* has been developed. Ammonia is used as the source of nitrogen, which makes nitriding more expensive than carburizing. It has three advantages, however: (1) It does not require so high a temperature (900°F. to 1100°F.), thus avoiding warping; (2) alloy steel,[1] or steel of higher carbon content, can be used because no heat-treatment is necessary; (3) the nitrided case is not only harder, but it is so resistant to corrosion that there are cases where it is replacing stainless steel.

[1] In fact, a certain percentage of alloy is necessary; this may be aluminum, aluminum plus chromium, or molybdenum; such alloying also adds to the expense.

Cyaniding

Cyaniding is an old process, still considerably used in spite of its extremely poisonous and rather messy nature. It is really a combination of carburizing and nitriding, for the formula of the cyanide ion is CN. It may be accomplished by immersing the part in a molten bath of a cyanide salt or by heating the part and then plunging it into the dry salt. A cyanide case is not deep, but it has the advantage of being quickly accomplished.

Flame and Induction Hardening

A simpler plan for obtaining a hard, wear-resisting surface, with a tough inner core, consists of heating the surface only and then quenching. Steel for this process must contain suf-

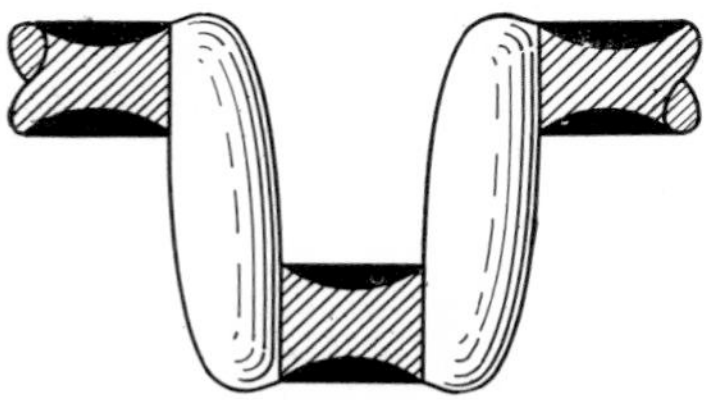

FIG. 55. Sketch of crankshaft; the black areas of the bearing cross sections indicate hardening as accomplished by induction.

ficient carbon to be hardenable. The heating may be done with a flame, but the most practical method is by *induction.* This is known as the *Tocco* process, and it lends itself extremely well to mass production of such parts as automobile crankshafts, where a tough interior is required but where the bearing surfaces must be wear-resistant. The operation is one requiring seconds only. Inductor blocks are snapped around the surface to be treated, and a high-frequency, low-voltage, high-amperage current is turned on. The frequency may be about a thousand cycles, and most of the heating is produced by eddy currents.

although magnetic hysteresis is also of assistance. Thus, two wasters of energy in electrical apparatus, which in that case are eliminated as much as possible by design, are in this case put to practical work (see page 181). Heating requires only a few seconds, and a water spray does the quenching. A crankshaft is partly sketched here with shaded areas hardened. The parts are machined nearly to size before hardening and ground to size after quenching.

DEFINITIONS[1]

Martensite is a microconstituent in quenched steel characterized by an acicular, or needlelike, pattern. It has the maximum hardness of any of the decomposition products of austenite.

Sorbite may be described as a late stage in the tempering of martensite, when the carbide particles have grown so that the structure has a distinctly granular appearance.

Spheroidizing consists in heating just under the lower critical point long enough for the cementite particles to attain a spheroidal shape.

Full annealing consists in heating iron alloys less than 100° above the upper critical, followed by cooling in still air.

Tempering is the process of reheating hardened steel to some temperature below the lower critical temperature, followed by any desired rate of cooling.

Austempering is the process of lowering the temperature of steel from hardening temperature to some temperature above that where martensite would form and of holding at that temperature long enough for complete transformation of austenite.

Nitriding is defined by "Metals Handbook" (1948) as a process of casehardening in which a ferrous alloy, usually of special composition, is heated in an atmosphere of ammonia, or in

[1] "Metals Handbook," pp. 1–16, American Society for Metals, Cleveland, 1948.

contact with nitrogenous material to produce surface hardening by absorption of nitrogen, without quenching.

Casehardening is heat-treatment or combination of heat-treatments in which the surface layer of an iron-base alloy is made substantially harder than the interior by altering its composition.

Carburizing is the process of adding carbon to the surface of iron alloys by heating the metal below its melting point in contact with carbonaceous solids, liquids, or gases.

Cyaniding is surface hardening by carbon and nitrogen absorption of an iron-alloy article by heating it in a molten cyanide, followed by quenching.

Martempering consists of rapid cooling of steel from above the critical range to a temperature just above the *Ms* temperature, holding at that point until surface and center temperatures are equalized, and cooling through the martensitic range in air. The purpose is to obtain a martensite structure without danger of distortion or cracking. An immediate tempering is advisable.

QUESTIONS

1. Which form of casehardening gives the most corrosion resistance?
2. Which one requires the least time?
3. Name some advantages and disadvantages of the nitriding process.
4. Explain the necessity in some cases for heating a steel more than once above the critical to put it in the proper condition.
5. What time would be required for complete transformation of austenite in an 0.95 per cent carbon steel at 600°F.?
6. Why is surface hardening by flame or induction not applicable to low-carbon steels?
7. Which is harder, a piece of steel tempered to a straw color or one tempered to a blue?
8. Why should a 0.20 per cent carbon steel be quenched from a higher temperature than an 0.80 per cent carbon steel?
9. How fast in degrees per second must an 0.89 per cent carbon steel be quenched in order to get past the knee of the S curve without transformation?
10. Explain how grain size increases if steel is held too long or heated too high above the critical temperature.

11. Why is spheroidized steel more easy to machine than annealed?
12. Give general trends in property changes
    a. With increasing carbon up to 0.85 per cent.
    b. With increasing speed of quench.
    c. With increasing tempering temperature.

REFERENCES

BULLENS, H. K., and METALLURGICAL STAFF OF BATELLE MEMORIAL INSTITUTE, "Steel and Its Heat Treatment," Wiley, New York, 1948.

"Metals Handbook," American Society for Metals, Cleveland, 1948. (See sections on Heat Treatment and Case Hardening.)

FIG. 56. G-E. conveyor-type scale-free electric hardening furnace, showing furnace, shaker loader, control equipment, and quench tank with oil-circulating cooling system. (*Courtesy S. W. Farber, Inc.,* and *General Electric Company.*)

Chapter 8
METHODS OF FORMING METALS

CASTING

A mold in its simplest form is simply a cavity in sand or metal into which molten metal is poured. But casting is a more complicated process than this definition indicates.

In the first place, the sand for a sand mold must be refractory (able to withstand heat), capable of holding its shape, and still porous enough to allow gases to escape.

The ideal molding sand is round-grained quartz sand with just enough colloidal (sticky) clay to bind the grains together. So important is the composition of sand that most foundries are now building up at least part of their sand from selected sands, clays, etc., which are known as *synthetic sands.*

Molds

The sand is rammed around the pattern in a flask, usually made in two sections, which is held in alignment by *pins* and *ears* *A*. The upper section is called the *cope* (*C*), and the lower one the *drag* (*D*). The pattern *P* is held in place on a plate *B* or on another device known as a *match.*

The drag is first rammed with sand (with the flask upside down); a board is placed over it, and then the flask is turned right side up and the cope rammed with sand. A hole is provided in the cope for pouring in the metal, and this is connected to the pattern cavity by runners and gates. Some powdered dry material is dusted on to prevent the sand from sticking to the plate. The cope is then lifted off and set to one side while the plate is removed and any cores that may be necessary are placed in position. The mold is then closed; the flask removed for the next mold; and the mold is set out ready to be poured.

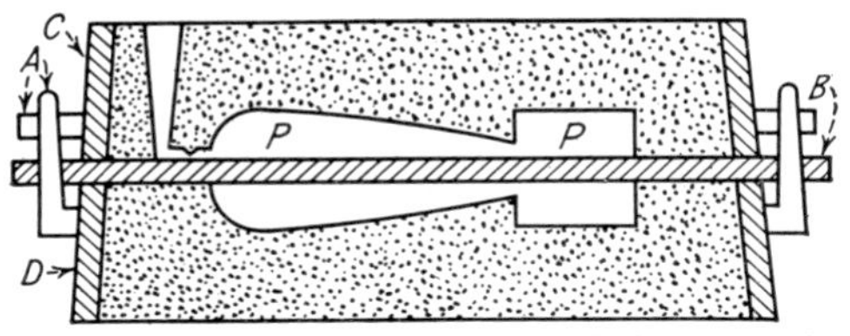

Fig. 57. Cross-sectional sketch of mold, in flask, with plate and pattern still in place.

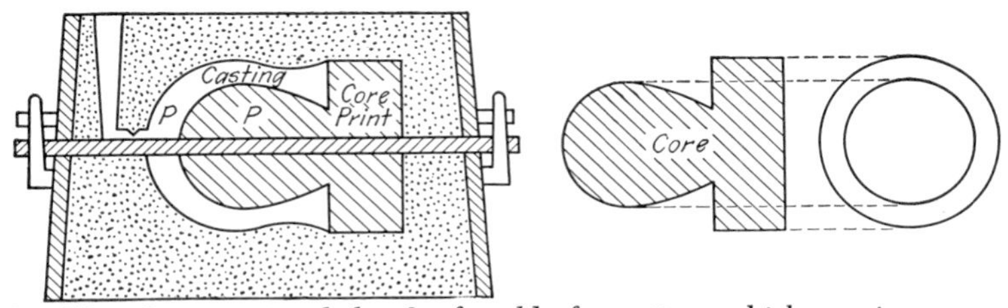

Fig. 58. Cross-sectional sketch of mold of a pattern which requires a core. The casting resembles a bell. Note that the pattern includes a section known as *core print*, which is not any part of the finished casting. Its purpose is to hold the core in place.

Properties of Castings

Most metals, on cooling, contract so much from their size as poured that the pattern must be large enough to allow for this.[1] Castings are also subject to internal *shrinks*. Such shrinkage is the result of the exterior of a casting solidifying

[1] Patternmakers use a *shrink rule*, which allows ⅛ in. per ft. for shrinkage.

first, while the interior is still molten. Then, when this interior solidifies and contracts, it has a tendency to form a cavity. Such interior unsoundness can be eliminated by feeding a reservoir of hot metal into the shrinking interior. Such reservoirs are called *risers* or *feeders* and often contain as much metal as the casting they feed.

Castings having both light and heavy sections adjoining are subject to cracking unless special precautions are taken. Even if they do not crack before being put into use, they may have *residual stresses* which will cause failures when an additional

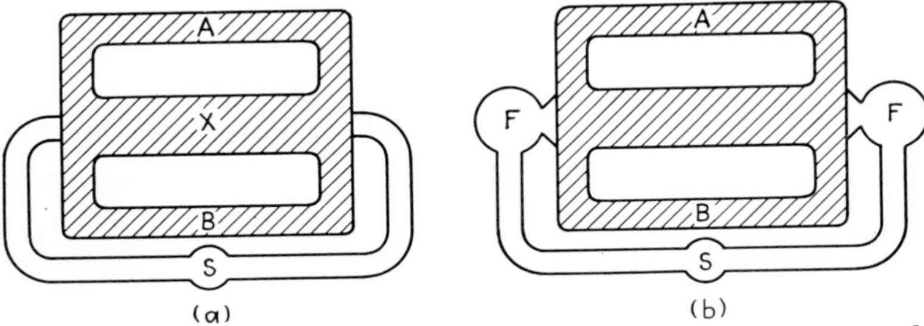

Fig. 59. Sketch showing a possible cause of unsoundness in a casting and methods of prevention.

stress is applied in use. The following experiment illustrates the effect of unequal cooling of adjacent light and heavy sections.

The arms *A* and *B* in Fig. 59 have about one-third the cross-sectional area of the center arm. Therefore they solidified first; so did the ends. These parts were rigid, while the center arm was plastic. When the center arm finally froze, it had to contract, and thus there was either a crack or a deep shrink at the point *X*. One method of preventing this trouble is to place feeders (*F*) where they will feed hot metal to the center arm as it contracts.

Besides such inequalities of contraction, there is a large

amount of expansion and contraction incident with iron or steel going through the critical temperature (as will be demonstrated in the experiment described on page 235). These facts emphasize the importance of uniform cooling in casting and quenching operations. (For by no means are these troubles encountered only in casting!)

Cores for shaping interiors of castings are made of dry, clayless sand, mixed with linseed oil, resin, or cereal as a binder and

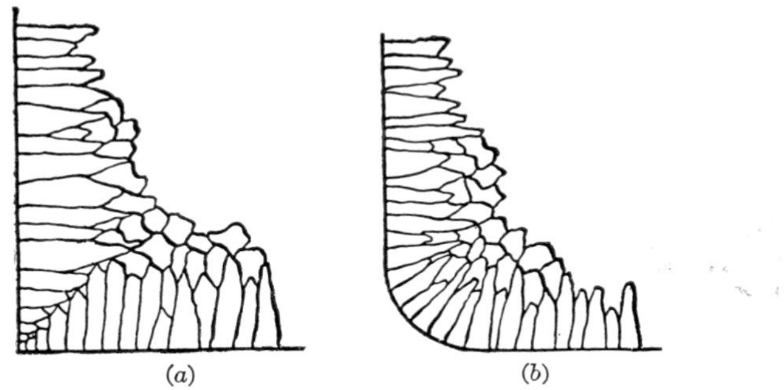

FIG. 60. (*a*) Plane of weakeness halving the angle in a sharp-cornered mold. Crystals grow inward in two directions at right angles to each other. (*b*) Elimination of weakness by rounding corner. Crystals grow inward from a curved surface and do not converge along a straight line. (*From G. E. Doan, "Principles of Physical Metallurgy."*)

then baked to render them hard. An illustration of where a core is necessary is shown in Fig. 58. Cores must have the following properties:

Green strength (to hold shape before baking)
Hot strength (to hold shape while metal is poured)
Permeability (to allow gases to escape)
Collapsibility (for easy removal from casting)

One source of weakness in large castings is coarse-grain growth (see Fig. 61). This trouble and its remedy were discussed under *normalizing*.

Centrifugal Casting

A special case of casting makes use of a circular permanent mold (usually of tungsten steel coated with adherent sand) into which just enough metal is poured to produce the casting. The mold is then whirled, causing the metal to shape itself outward against the mold. The result is a firm casting, free from shrinks or blowholes. Cast-iron water and sewer pipe is made by this method, as are many bearings. Another method of ensuring solid castings is to cast in a steel mold by hydraulic pressure (used mainly with brass and bronze).

FIG. 61. Columnar grains perpendicular to cooling surface in a chilled casting (see also Fig. 60). (*From Doan and Mahla, "Principles of Physical Metallurgy."*)

These molds are provided with a well that has an opening at the bottom connected with the mold proper. The operator fills the well with molten metal from a hand ladle, and a plunger worked by hydraulic pressure immediately descends into the well and forces the metal into all parts of the mold cavity.

Forging

As was hinted above, the grain size and shape are not always what is desired in a casting. It is obivious that if the metal is hammered, pressed, or rolled while in a plastic stage, it will be denser and its grain size and shape improved. Many parts formerly made as castings are now forged—shaped between dies or hammered into shape (see Fig. 62). See page 141 for a discussion on grain structure of forgings.

Method	Principle of operation	Action on metal
Forging Hammering Pressing	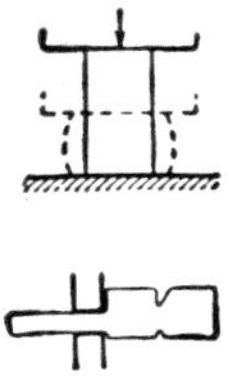	Pressure on unconfined metal
Die forging	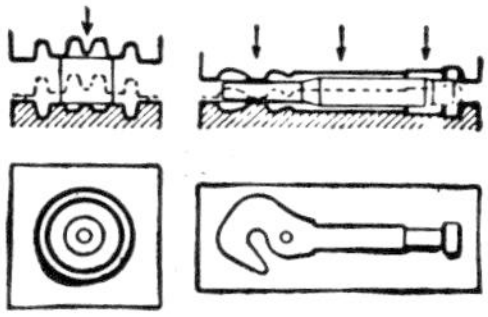	Forces metal into a die
Extrusion	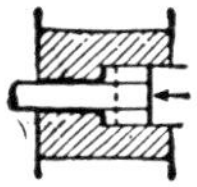	Squirts metal through a die
Piercing	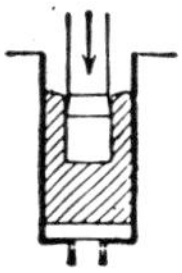	Pressure exerted by mandrel on confined metal
Wire drawing	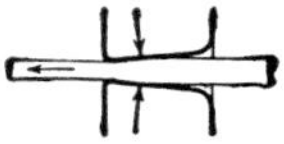	Wire or rod drawn through die
Tube drawing	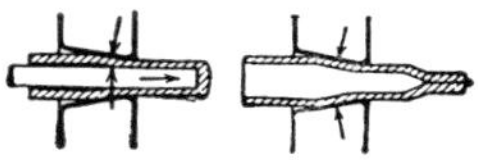 (*a*) (*b*)	(*a*) Reduction in wall thickness between die and plug (*b*) "Sinking" or reduction in diameter

FIG. 62. Several methods of forming metals. (*From G. E. Doan, "Principles of Physical Metallurgy."*)

Rolling

More steel is shaped by this method than by any other. All rails and structural shapes and all flat-rolled products are so manufactured. Steel for rolling is cast into ingots as it leaves the furnace, or converter. These ingots are simply large elon-

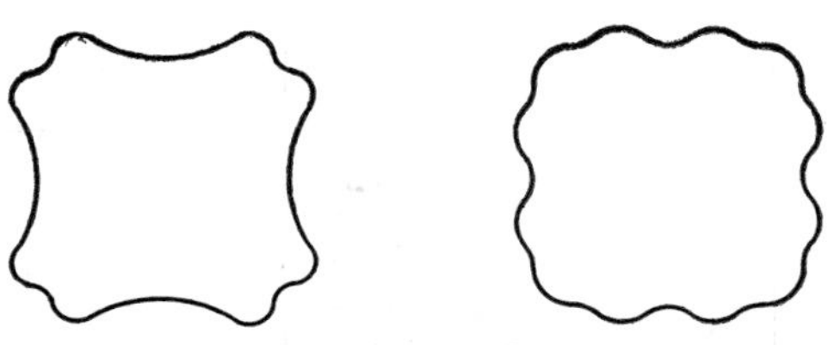

FIG. 63. Ingot outlines. (*Courtesy Gathmann Engineering Company.*)

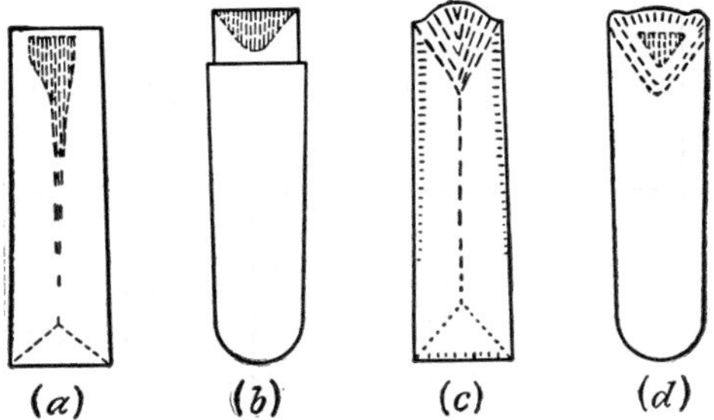

FIG. 64. Ingot-mold designs. Notice effect of "hot-top" in (*b*). Another improvement in (*b*) over (*a*) is that it is cast "big end up," thus greatly assisting in escape of gas. (*a*) and (*b*) represent killed steels, while (*c*) and (*d*) are not thoroughly degasified. However, (*d*) contains fewer gas holes than (*c*) because cast "big end up." Notice rim on (*d*). (*Courtesy Gathmann Engineering Company.*)

gated castings, often weighing many tons. They are subject to the same imperfections as castings. If an ingot were poured into a mold of square cross section, the internal structure would resemble Fig. 60*a* because of granular growth inward from a chilled surface. This accounts for ingot designs, such as in Fig.

63. The weak arrangement of crystals is not the only source of weakness, for as the grains are solidifying and growing inward, they are pushing ahead of them impurities that are not particularly soluble in the metal, thus concentrating these impurities at the point that solidifies last—the center. This center weakness is, of course, very much accentuated at the top, be-

FIG. 65. Stripping alloy-steel ingots from ingot molds. (*Courtesy Bethlehem Steel Company.*)

cause hot metal descends from above to fill the lower portion of the center. Thus there is formed a cavity at the top known as a *pipe,* which must be discarded before manufacture and which accounts for a portion of the steel scrap, which the open-hearth furnace must melt. In extreme cases, with poor design, this could amount to one-third of the ingot (see Fig. 64*a*). Figure 64*b* shows one method of combating this trouble. It is known as the *hot-top* ingot. The top portion (smaller in cross section than the ingot) is made of a refractory material, such as brick, which conducts heat very slowly, thus furnishing a well of hot molten metal, which feeds down and eliminates most of the *pipe.* Another advantage of the hot-top ingot is that it is cast *big end up* instead of *down,* thus assisting the escape of gas.

"Rimmed Steel"

Molten steel can hold in solution much more gas than solid steel. One class of low-carbon steel makes use of this property to eliminate pipe. Because the surface solidifies first, it expels gas toward the center, causing the center to puff up, thus eliminating pipe. After solidification, such an ingot consists of two distinct zones: an outer rim of comparatively pure steel (hence the name *rimmed steel*) and a core containing scattered blowholes and a minimum amount of pipe. This steel is in demand for the production of sheets because of its excellent surface. The blowholes are welded shut in the rolling operation. Steels that rim best contain not more than 0.15 per cent carbon, while steels of more than 0.30 per cent carbon cannot be made to rim at all.

Often steels are *killed,* meaning that the gas is removed by the addition of suitable *deoxidizers* in the ladle, such as silicon, manganese, and aluminum. "These reduce the oxygen content of the metal to a minimum so that no reaction occurs between carbon and oxygen during solidification. *Semikilled* steel is incompletely deoxidized in order to permit of evolution of sufficient carbon monoxide to offset solidification shrinkage."[1]

The operations of rolling will not be described here because they can be so well illustrated by easily obtainable motion pictures. It is necessary here to emphasize that rolling or any mechanical working changes the grain structure from granular

[1] Definitions from "Metals Handbook," American Society for Metals, Cleveland, 1948.

and irregular, like granite, to a more or less fibrous structure which has more strength (at least in the direction of flow) than the original (see Fig. 66). This subject will be further discussed in Chap. 9, on Grain Structure of Metals.

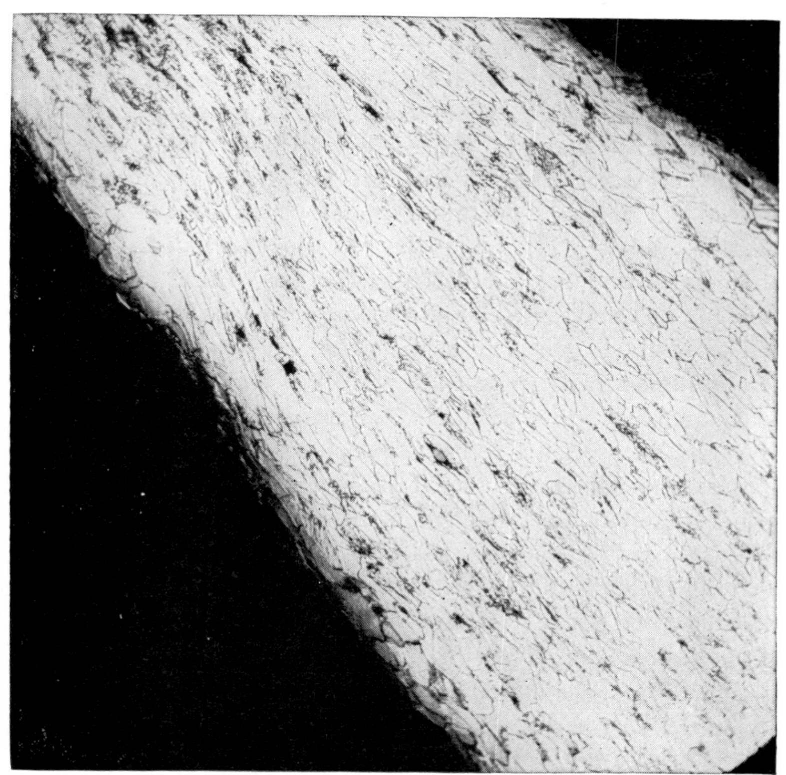

Fig. 66. Grains elongated by forging or rolling. Magnification 100 diameters. Notice fibrous structure. (*Courtesy General Electric Company.*)

Another fact that must be noted in this connection is the great difference between *hot-* and *cold-working,* or rolling. "Shapes" such as rails and structural shapes are formed by hot-rolling. The purpose of rolling is not concerned entirely with the production of definite shapes. An equally important function is improvement in texture of the metal.

It is easy to see that hot-working will break up the usual large-grain structure of the ingot into a finer grained texture; however, as will be further explained in Chap. 9, there is a tendency for these fine grains to merge. This tendency increases with temperature. Therefore, with fine-grain structure in view, the rolling or other hot-working, such as forging, should continue down to a temperature where "recrystallization," because of rigidity of atoms, cannot occur. In fact, the definition of *hot-working* as given in the ASM "Metals Handbook"[1] is: "Plastic deformation of metal at such a temperature and rate that strain-hardening does not occur. The lower limit of temperature for this process is the recrystallization temperature."

The same reference gives the definition of cold-working as: "Deforming a metal plastically at such a temperature and rate that strain-hardening occurs." The upper limit of temperature for this process is the recrystallization temperature. Thus, the strain-hardening effect of cold-rolling is seen to be one of the principal differences which distinguish this process from hot-rolling. Cold-working crushes and distorts the grains to such an extent that hardness is increased and ductility is decreased in proportion to the amount of deformation. If the process is continued too far, it results in such an internal strain that the metal breaks. This strained condition is well shown in Fig. 79, while Fig 80 shows how heating up to a recrystallization temperature can eliminate the strained condition of the crystals.

When cold-working or rolling operations have to be carried on to the extent of causing the metal to become brittle, it must be put in condition for further reduction by heating to a temperature at which the atoms become mobile enough to re-

[1] "Metals Handbook," pp. 4 and 7, American Society for Metals, Cleveland, 1948.

crystallize. This temperature is not usually as high as a full anneal, but the operation is usually referred to as an *anneal.* See the description of wire drawing in Chap. 9.

Cold-working is resorted to for various reasons. In rolling, it is used to give a smooth surface, to get precise dimensions, and also to obtain desired hardness. (Hardness so obtained is often referred to as *temper.*) In cold forming, such as in automobile fenders, refrigerator cabinets, etc., the cold method is used because, with the class of steel used, there is no necessity in using heat to get the desired plasticity. Several types of rolls are especially designed for cold-rolling. In one, the Steckel mill, the strip is drawn through the rolls by tension on the take-up reel instead of by applying power to the rolls themselves. This mill is four high; the two center rolls through which the sheet is pulled are well polished and of small diameter, while the two outer rolls are massive and serve to stiffen the small inner rolls. The operation is comparable to wire drawing, with rolls substituted for the dies.

Welding and Piercing

The manufacture of tubing is described because it is a special case of rolling and involves either welding or piercing. There are three methods of forming tubing: the butt welding, the lap-welding, and the seamless. Butt welding is the simplest. As the name indicates, the material is welded at one point in its circumference by simply butting together the two edges of a long narrow strip, known as *skelp.* The skelp is heated to welding temperature, passed through a *welding bell* (see Fig. 67), and then rolled to correct size. Butt welding is done on pipe, ⅛ in. to 3 in. inside diameter.

Lap-welding results in a stronger weld. In this case the skelp is beveled on its edges. It is partially formed by drawing through a bell, as before; then, after being reheated to welding

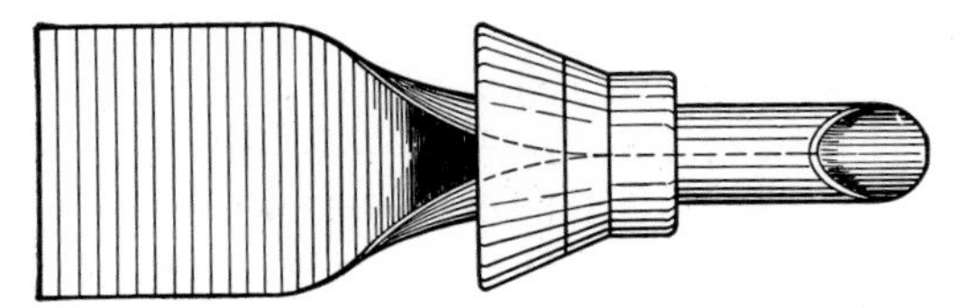

FIG. 67. Sketch showing how the flat skelp is formed into a tube by drawing through the welding bell. (*Reprinted from E. E. Thum, Editor, "Modern Steels," American Society for Metals.*)

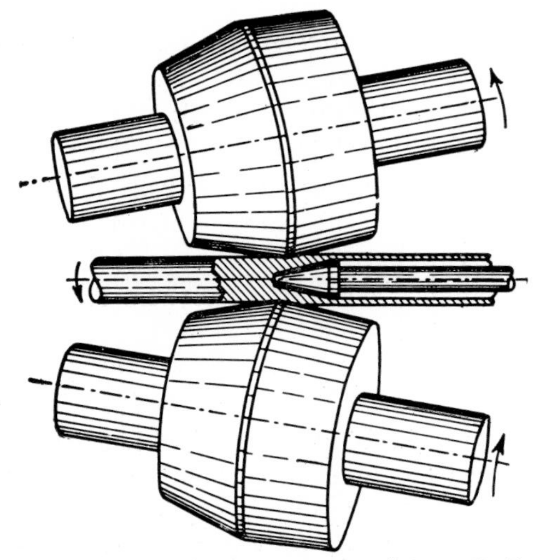

FIG. 68. Piercing seamless tubing. (*Reprinted from E. E. Thum, Editor, "Modern Steels," American Society for Metals.*)

temperature, this tube is caught between a pair of rolls and rolled over a manganese-steel mandrel so that the beveled edges are welded by pressure.

Piercing

This pipe is not strong enough to withstand pressures such as are encountered in hydraulic presses, etc. Pipe for such uses is made seamless by simply piercing a hole through a rod at white heat while it is being pushed forward against the tungsten-steel mandrel by special rolls. These rolls are set up at an angle of about 10 deg. to the line of rolling, as Fig 68 attempts to show. Their shape is such that they grip the billet and push it forward.

Drawing (wire and tubes)

The usual procedure in making wire is to draw cold hot-rolled wire rod through a succession of dies to the required diameter, the motive power being the reel upon which the wire is wound. Sometimes an intermediate anneal between draws is necessary (see Chap. 9). Pipe is also often reduced in diameter by being pulled through dies.

Extrusion

Extrusion is a method of forming which is constantly finding new applications. One of its original applications was in the forming of lead pipe and lead cable sheaths. Figure 70 sketches an apparatus for forming the latter. Lead is forced out at a temperature of 425°F. and surrounds the cable with a very tight-fitting cover. In the case of cable, no mandrel is necessary, because the cable forms the core. Lead is a very easily formed metal, but by improvements in apparatus and with the use of higher temperatures, even copper pipe is now extruded. Collapsible tubes such as tooth-paste containers were formerly extruded from tin, but on account of the scarcity of that metal, aluminum was substituted to a great extent. It was found that this metal can be easily extruded if it is of extreme purity, say 99.7 per cent, or better. Magnesium is another metal which is extensively extruded. Extrusion lends itself well to the production of rods, bars, and even complicated shapes which could not be made in long lengths in any other way.

Method	Principle of operation	Action on metal
Stamping	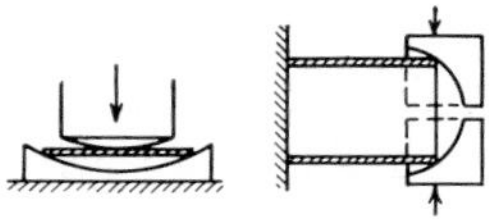	(a) Curving plates (b) Curving ends of cylinder
Cupping Flanging	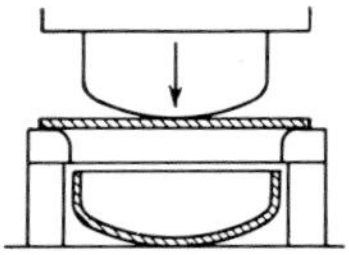	Bending edges of disk
Deep drawing	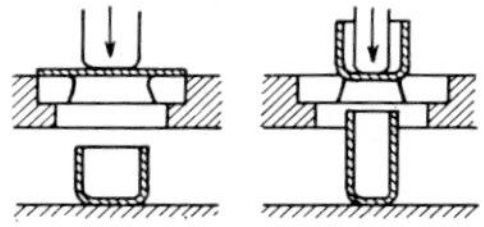	Stretching of walls (cartridge manufacture, automobile bodies)
Bulging	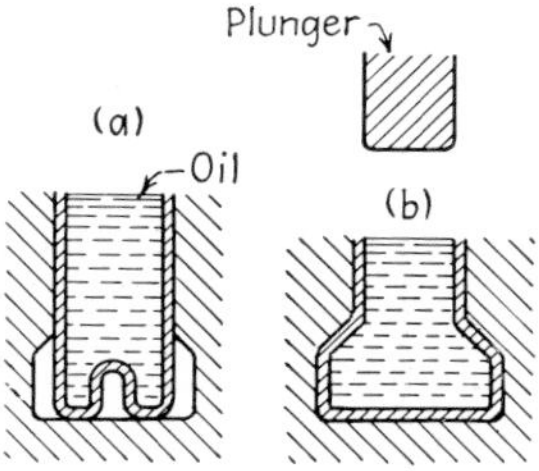 Plunger (a) Oil (b)	Lower part expanded by oil pressure
Rolling of sheets, beams and rails	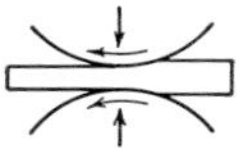	Stretching by rolling
Piercing (Mannes-mann)	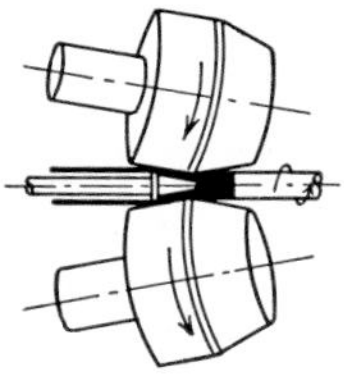	Rolls and mandrel rotate in same direction

FIG. 69. Additional methods of forming metals. (*From G. E. Doan, "Principles of Physical Metallurgy."*)

Stamping, cupping, flanging, and *deep drawing* are self-explanatory from cuts in Fig. 69. Note that deep drawing requires two or more operations with intermediate annealing. It would be well to call attention to the fact that the word *drawing* has two separate meanings in metallurgy. In its usage in connection with forming of metals, it means a drawing through dies, as in the case of wire manufacture, or, as mentioned above, forming in which the metal is subjected to tension in the dies.

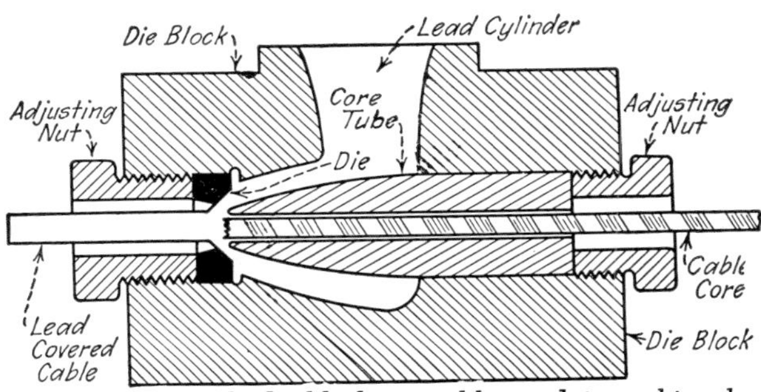

FIG. 70. Section through die-block assembly used in making lead cable. (*Reprinted from "Metals Handbook," American Society for Metals,* 1939.)

The other meaning of the word is synonymous with "tempering" in heat-treatment.

Note the cut designated *bulging.* There are two methods of forming containers whose lower part is bulged out larger than the upper part. One method is illustrated in the cut: The shape at (a) is placed in the die and filled with oil. The plunger then comes down and by pressure transmitted through the oil causes the bulge shown at (b). Of course the die must be made split so that it can be removed. Rubber is also used instead of oil as the bulging medium.

The other method of forming such shapes is by *spinning.* Many household articles such as water pitchers and tea kettles are made of "spun" copper, aluminum, or even stainless steel.

The process starts by clamping the circular disk or cup in a lathe and working the outside edge inward and forward, much as a potter would form the neck of a vessel, such as a water bottle, on his wheel. The tools used both on the outside and inside are some kind of roller, although maple wood can be used in some cases. A blowtorch flame is sometimes employed to keep up the proper temperature.

Fabricating by Welding

This process should be mentioned in connection with the forming of metals because many shapes are built up by welding. Torch welding should be done with a proper understanding of the effect of oxidizing and reducing flames, especially on steel. Too much oxygen will cause the weld to be brittle, because of iron oxide. Too much carbon has almost the same effect, as it results in higher carbon in the weld than in the rest of the metal, forming high-carbon steel, which is brittle.

The carbon content of the weld can be controlled in two ways: First, the flame, if oxyacetylene, can be made oxidizing (using excess oxygen) or reducing (using excess acetylene); second, if oxidation is expected, a welding rod with a higher carbon content than that of the work is often used.

Besides such chemical aspects, there are also physical and thermal problems to consider, and so a weld can be a very complicated job.

After a study of Chap. 5, one can realize that a welded joint has been subjected to every phase of heat-treatment from molten metal down through transformations in the critical range and their attendant expansions and contractions, and finally the contraction on cooling down to ordinary temperatures—sometimes all within the space of an inch or less.

To relieve such stresses, welds should be either peened or

annealed. For example, take the electric welding of a band saw. The ends are clamped into jaws that hold the ends in alignment and serve as electrodes. The ends are fused together by a current, but because the cold electrodes are so close, the joint chills very quickly, becoming hard and brittle. In other words, it is *quenched.*

To relieve this brittleness, the electrodes are again clamped in, but at a greater distance apart. The current is then turned on for a short time, just long enough to bring the weld and the metal, for a short distance on each side, up to a cherry red. The joint and adjacent metal cool more slowly this time because they are not heated so high and because a larger amount of metal is heated.

Thus it can be seen that welding involves much more than merely joining two metals together, as would be the case in soldering. The small molten bath of metal in a weld can be compared to a miniature open-hearth furnace in its oxidation and reduction possibilities; and after the bath of metal solidifies, it is subject to all phases of heat-treatment, either controlled or uncontrolled.

Machining

This is an important method of shaping metal parts and especially of finshing them to dimensions. Ease of machining is one of the principal factors to be considered in the preparation or selection of metal that must be so finished. The carbon particles of gray cast iron and malleable iron make these irons very easy to machine. Many steels must be in the annealed condition for machining, as has been discussed under *heat-treatment.* The special steels that are used in making the cutting tools will also be considered. Generally, any finishing after hardening must be done by grinding.

Powder Metallurgy

While this is not really a new method of forming metals, its applications have grown to such an extent recently that it is considered a new process in many fields. The powdered metal is forced into strong molds under hydraulic pressure, which is so great that the article thus compressed will hold its shape. The

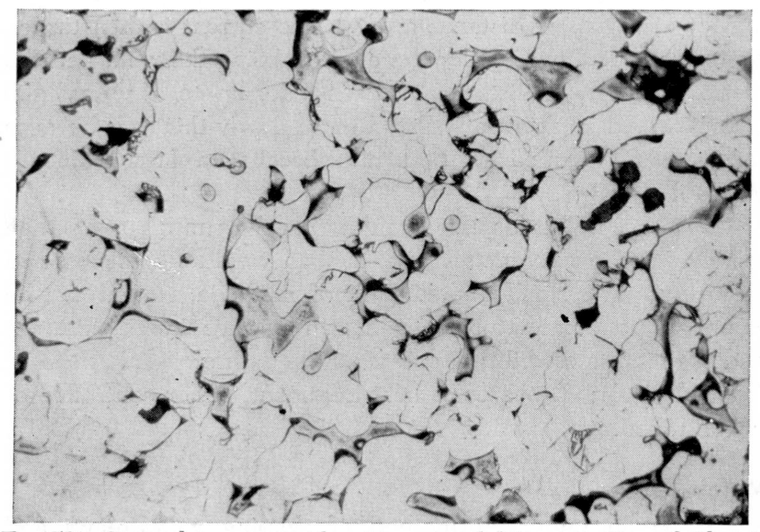

FIG. 71. Sintered iron copper bearing material (*Courtesy General Electric Company.*)

article is then "sintered" (heated to incipient fusion) and results in a solid piece of metal, not quite as dense in most cases as if it were poured or if it were shaped by working, but strong enough for many purposes.

Some of the small pinion gears for automobile water pumps are made in this way. A great advantage of this method is that parts can be finish-formed to tolerances of 0.001 in. perpendicular to direction of pressure and 0.005 in. parallel to pressure. Finish

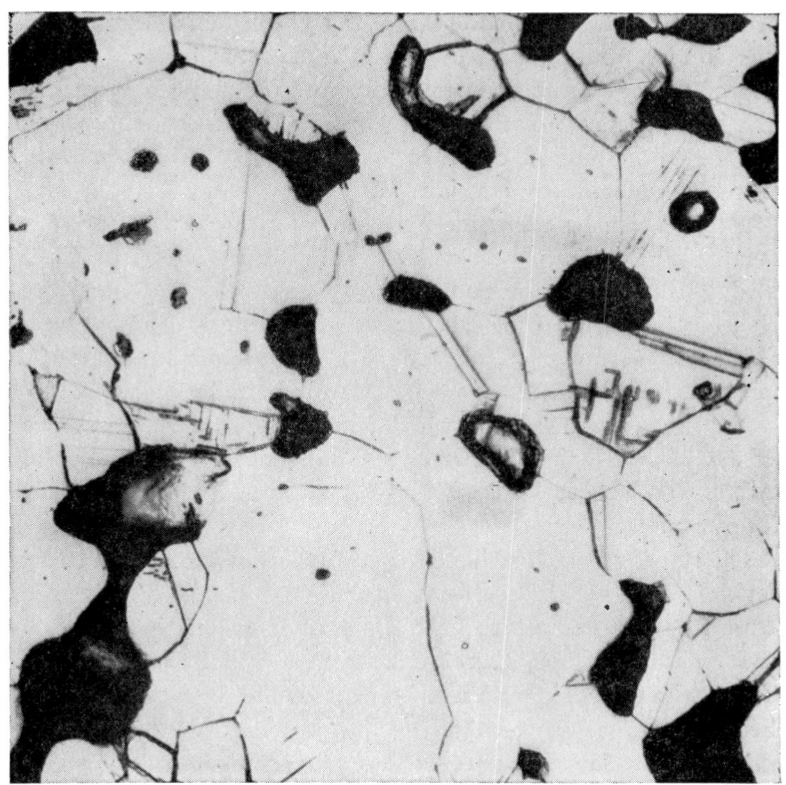

FIG. 72. Photomicrograph of "oilite," a bearing bronze manufactured from powdered metals by the Amplex Manufacturing Company, a subsidiary of the Chrysler Corporation. Magnification 750X.

is limited only by smoothness of die. A great deal of machining is thus avoided.

Many alloys can be made by using a mixture in the proper proportions of the powdered component metals, and some alloys that could not be combined by any other method are thus possible to produce. For example, a combination of tungsten and copper makes an excellent alloy for electrical contacts, but if an attempt were made to produce this alloy by melting, the copper would vaporize before the tungsten melted. Sintering of the powdered mixture results in an alloy which combines the heat resistance of the tungsten with the high conductivity of the copper. Sintered carbides for cutting tools are a product of powder metallurgy (see Chap. 10).

Porosity can be overcome by a second application of pressure while still hot from the sintering, but porosity is often desired, and it can usually be controlled between 5 and 50 per cent. Self-lubricating bearings, such as in refrigerators, electric clocks, door hinges, and a multitude of similar applications, are so produced, with the lubricant being usually oil or graphite.

When military and commercial projects caused a great deal of machinery, such as airplanes and bulldozers, to be used under Arctic weather conditions, oil-impregnated bearings prevented a lot of trouble.

On starting such machinery, if the shaft becomes heated for lack of oil, it will in turn heat the bearing, which will exude oil from its porous structure (like the wick in a kerosene lamp) to the bearing surface, thereby creating a full film lubrication. Then when the sluggishness of the regular oiling system has been overcome by warming up, the oil will flow back into the pores ready for the next cold start. Porosity is sometimes obtained by mixing in combustible material when making up the powder.

One of the earliest applications of powder metallurgy was in the production of the tungsten wire that is used in electric-lamp bulbs. Tungsten is one metal which because of its high melting point really "cannot be melted *in* anything." There is no refractory which can withstand the 6,000, or more, degrees required for bringing tungsten into its liquid state. The method of manufacture is as follows. Tungsten ore is treated by hydrogen, and the resulting metal is reduced to a powder. This powder is compressed into bars about 1 in. square and 18 in. long, under such pressure that the bars can be taken out of their mold and stood upright. These bars are then heated by their own resistance to the passage of an electric current through them until they are almost on the point of melting. They are then swaged and rolled (with gas flames playing on them) until they are small enough to be drawn through dies into wire. These dies are actual diamonds, and the drilling of holes through them by means of diamond dust and rapidly revolving spindles is one of the interesting side lights in tungsten wire manufacture. Molybdenum, columbium, and tantalum metals are produced in a similar manner.

Among the various methods of producing metal powders are grinding, spraying, electrolysis, precipitation, and reduction. It might be mentioned that the oxides of iron, nickel, tungsten, copper, and cobalt can be reduced to form powder by reducing in an atmosphere of hydrogen.

QUESTIONS

1. Can you differentiate between contraction and shrinkage in castings?
2. Name some inherent weaknesses in castings and suggest remedies.
3. Why may a forging be superior to a casting?
4. What is an ingot?
5. What keeps the metal hot in a hot-top ingot?
6. Why is rimmed steel preferable for sheet steel?
7. Describe four methods of forming tubing.
8. Name three different welding methods mentioned in the chapter.
9. Why should sharp corners be avoided in castings?
10. Can you tell why welding rods are often coated?

REFERENCES

AMERICAN FOUNDRYMEN'S ASSOCIATION, "Cast Metals Handbook," Chicago, 1944.

AMERICAN IRON AND STEEL INSTITUTE, "The Picture Story of Steel," 350 5th Ave., New York.

BEGEMAN, MYRON L., "Manufacturing Processes," Wiley, New York, 1947.

Clapp, W. H., and D. S. Clark, "Engineering Materials and Processes," International Textbook, Scranton, Pa., 1949.

Naujoks, Waldemar, and Donald C. Fabel, "Forging Handbook," American Society for Metals, Cleveland, 1939.

Pearson, C. E., "The Extrusion of Metals," Wiley, New York, 1944.

Rossi, Boniface E., "Welding and Its Applications," McGraw-Hill, New York, 1941.

Wulff, John, Editor, "Powder Metallurgy," American Society for Metals, Cleveland, 1942.

Chapter 9

GRAIN STRUCTURE OF METALS

A specimen being prepared for microscopic examination must be first polished to such a degree that no scratches can be detected by the microscope. Usually the surface appears perfectly plain at this stage, except for scattered inclusions and nothing can be ascertained as to its structure.

The next step is to etch this polished surface by some reagent, usually an acid, which acts with varying intensity on the different constituents. This will be explained more fully in Chap. 12, but what we wish to mention at this time is that this etching acts more vigorously on grain boundaries than on the grains themselves and thus outlines all the separate grains. This fact is evident in most of the photomicrographs.

Grains

Just what causes metal to crystallize into individual small grains instead of being of uniform structure throughout? Both upon solidification and in the various transformations, crystals start at many different places. Each nucleus of a crystal as it starts to solidify gets started in a different direction from its neighbors, for it must be understood that atoms arrange themselves in space in a very orderly geometric pattern.[1] Mention

[1] Spoken of as a *space-lattice.*

was made in Chap. 5 that iron atoms, for instance, arrange themselves in some kind of a cubic pattern. Let us represent this in as simple a manner as possible by dots arranged in squares.

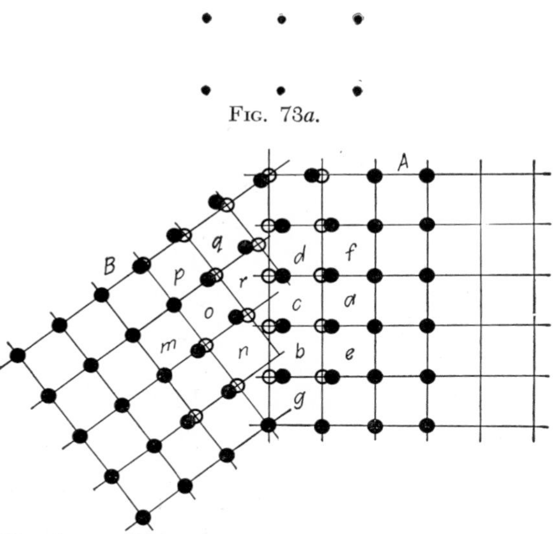

Fig. 73*a*.

Fig. 73*b*. Showing conflicting space-lattice and consequent atomic strain at grain boundaries due to differences in orientation in adjacent grains. Circles indicate natural positions of atoms if there had been no influence by adjacent orientation.

What was meant in the last paragraph in regard to atoms getting started in different directions could be illustrated (see Fig. 73*b*) by saying that the first four atoms to settle down in crystal *A* were those around the cube *a*. Not far away four others were settling down as the cube *m*. At the time these arrangements were being made, there was no connection between *a* and *m*; the atoms intervening were still in a molten condition. As soon as *a* started to solidify, there was a rapid settling down of *b*, *c*, *d*, *e*, *f*, *g*, etc., in line with *a*. Similarly, when *m* settled down, *n, o, p, q, r, s,* etc., arranged themselves in line with it. Soon the two patterns met.

Grain Boundaries

Those atoms which were last to arrange themselves found themselves influenced by two different alignments. The tendency of atoms to orderly arrangement is very strong. Those at the boundary had to arrange themselves as best they could; but not only were they under a strain, they put those for several atoms back under strain. The metal here at the boundary had different qualities from that in the interior of the grain. Being under strain, it was harder.[1]

That is the explanation for the existence of so many separate grains. It is also the explanation for grain boundaries being stronger than the grains themselves.

Finally, it also explains why fine-grained steels are stronger and tougher when cold than coarse-grained steels. The smaller the grains, the more grain boundary per unit volume of steel. The word *cold* was specified above because at high temperatures the grain boundaries are the weakest part of the metal. The higher the temperature, the more mobility the atoms have. The boundary atoms are no longer under strain and thus become the weakest part of the structure. The photomicrograph (see Fig. 74) shows how the grains can be separated by heat. The very forces which caused the strength in the grain boundaries when cold now shove the grains apart and allow them to align themselves with their neighbors.

Forging or rolling operations which are carried on at fairly high temperatures should never be stopped at those tem-

[1] There is one more factor that enters into the strength or weakness of grain boundaries. If any impurities were present when solidification started, such stranger atoms were probably not wanted and were pushed ahead into the grain boundary. In some cases these stranger atoms weaken the boundary; in others they may strengthen it.

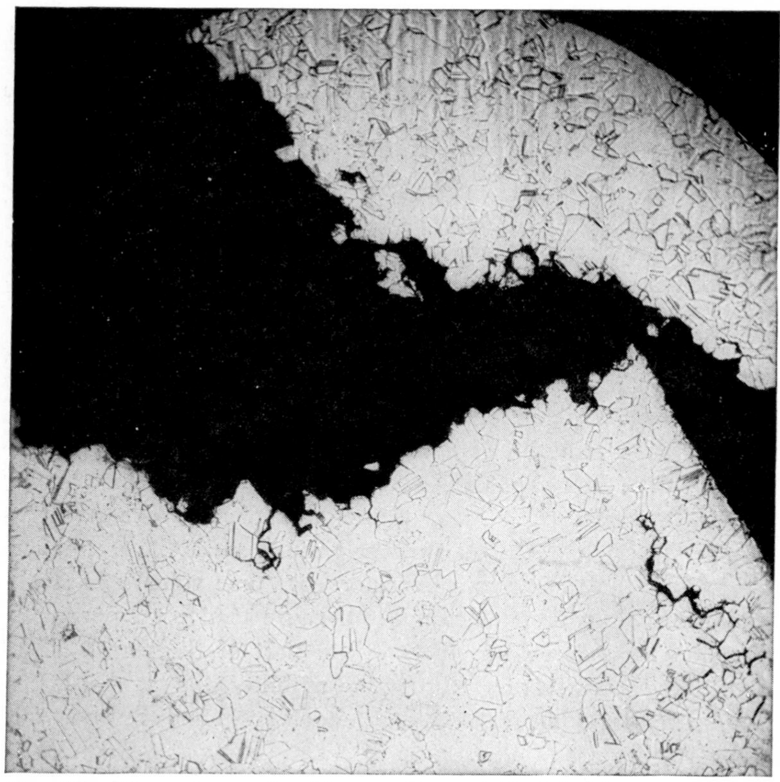

FIG. 74. Photomicrograph of brass tubing which failed because of overheating. Notice separation at grain boundaries. (*Courtesy General Electric Company.*)

peratures but should continue all the way down to the critical range in order to prevent grains from coarsening. The closer to the lower limit of recrystallization the finishing is done, the more of the grain refinement will be retained. Finishing between the A_{r1} and A_{r3} temperatures results in a retention of a banded structure in the product, due because of the lengthening of grains and the growing inability of the grain structure

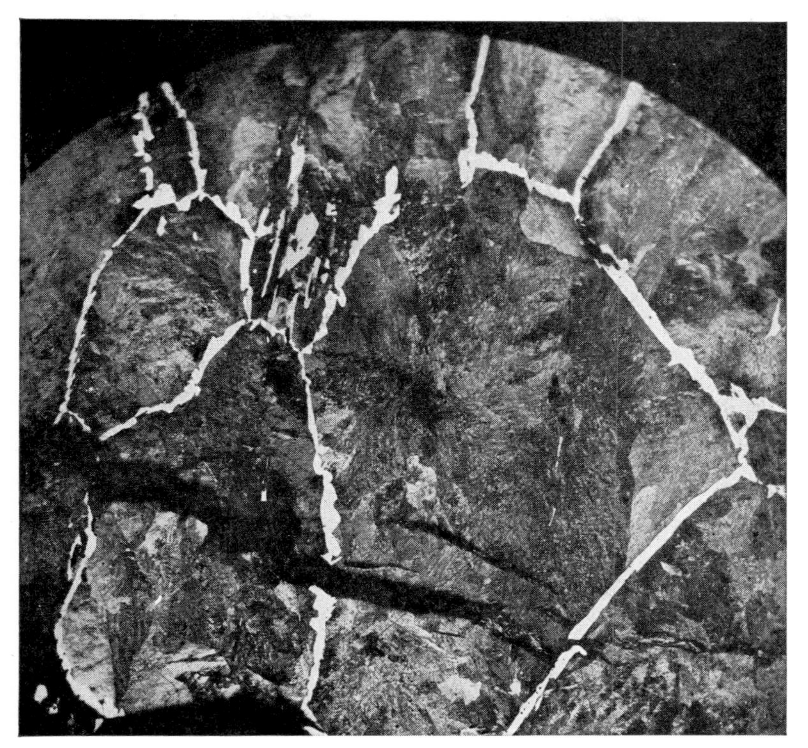

FIG. 75. Photomicrograph taken at the fracture of a test bar, showing that fracture occurs within the grains rather than at their boundaries. (*Courtesy General Electric Company.*)

to readjust itself. The recrystallization temperature does not always coincide with the lower transformation temperature but is related to it. Speed of working can lower the temperature, and, of course, alloys affect the temperature because they affect the transformation temperatures. Similiarly, two factors affect the temperature at which recrystallization occurs when annealing cold-worked metals. The more drastic the deformation by cold-working, the lower the recrystallization temperature,

This is because of the greater amount of energy locked up in the grain boundaries. Within limits the temperature at which a cold-worked metal begins to recrystallize is lowered by increasing the time of anneal.

COLD-WORKING

Let us picture what takes place when a metal is deformed cold. Figure 75 was taken at the fracture of a test bar under

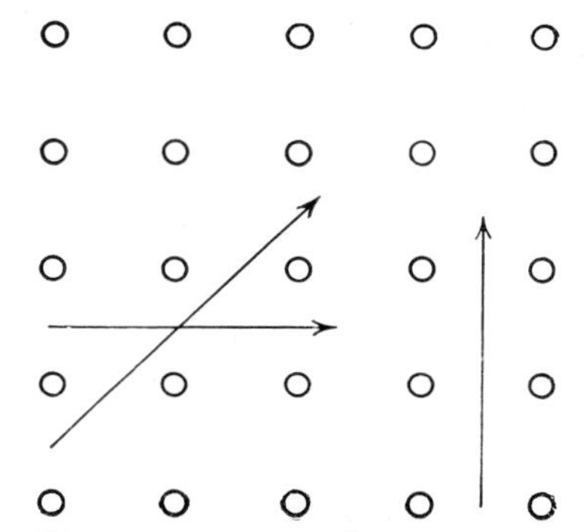

FIG. 76. Arrangement of atoms in squares.

tension. Note that fracture occurred within the grain instead of along grain boundaries. Grain boundaries are the stronger, and cold-working increases their number.

What happens inside the grain can be illustrated by referring to the arrangement of atoms in squares, as in Fig. 76, which, while it may not fit the case exactly, at least has the advantage of simplicity. Suppose Fig. 76 represented trees in an apple orchard; anyone walking around the orchard would come to different points (represented by arrows) at which he could see right through the orchard. The weakest part of a crystal or grain is along planes where there are no atoms in the way of

FIG. 77. Photomicrograph showing slip planes developing in the grains of a strained metal. (*Courtesy General Electric Company.*)

allowing a slip. Such planes are called *slip planes* (see Fig. 77). There is a great deal of microscopic proof that cold-working begins this way. The grains become elongated by continual slipping along planes, a process somewhat analogous to the overlapping of a pack of cards, as shown in Fig. 78.

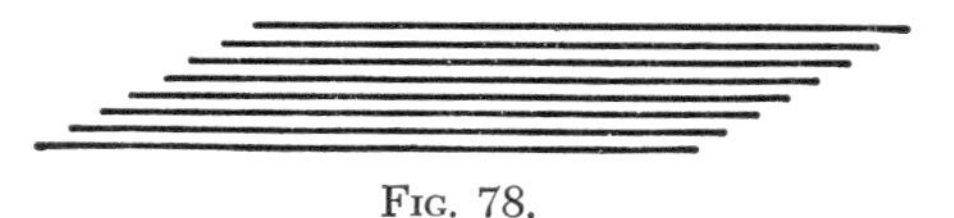

FIG. 78.

The grains also become broken up into smaller grains. This is the explanation for the fact that so many metals can be hardened by cold-working. Of course, there are a few metals,

Fig. 79. Photomicrograph of a badly strained piece of brass.

such as lead and tin, the recrystallization temperatures of which are down at or below room temperature. Any working at room temperature would correspond to *hot-working* for such metals. Figures 79 and 80 show the effect of overstraining on grain structure and the recovery by annealing, or recrystallization.

FIG. 80. The same piece, recrystallized by proper heating. (*Courtesy General Electric Company.*)

Cold-working is practiced in industry a great deal for special purposes. For instance, copper wire for transmission lines is cold-drawn to give it sufficient strength to prevent sagging. In this case it is out of the question to use an alloy as a hardener because any element, even in the smallest proportion, would decrease conductivity. In fact, all wire is cold-drawn, and great strengths are obtained; so much so that a wire cable can be stronger than a rod of the same size. Generally, several drawing operations are needed to get wire down to size, combined with two or three annealings whenever the wire becomes too brittle.

The manufacture of steel wire is such a good example of grain deformation by cold-working and subsequent heat-treatments for recrystallization that it will be briefly outlined here. In case a soft wire is desired, steel of low carbon content—say 0.10, is used. The original ¼-in. bar, having a tensile strength of 68,000 p.s.i and a ductility represented by 25 per cent elongation, is pulled twice through dies, which increases its length 80 per cent. Its tensile strength has now increased to 150,000 p.s.i., and it is so brittle that its elongation is only about 1½ per cent. No further drawing is possible for the wire would break; and so a softening operation is restored to which consists in heating to somewhere between 1020°F. and 1200°F. The sketches in Fig. 81 show what has taken place in the microstructure of the metal so far. The original structure would be as in (*a*)—ferrite grains interspersed by a few pearlite grains. Drawing would elongate the grains, as in (*b*). The reheating would give the ferrite grains a chance to resume some of their original shape, as in (*c*). They were in the condition of a stretched piece of rubber, and, given an opportunity, they came back to normal. Inasmuch as the temperature of the heat-treatment was below the critical range, the pearlite grains were affected very little. Such a heat-treatment is known as *process anneal.* It practically restores the original properties, and the wire is ready for further drawing. Such wire is finished by some degree of anneal after the final draw.

In case a stronger and less pliable steel wire is desired, a higher carbon content is used, say 0.50 per cent. In this case the pearlite grains are much more numerous (see Fig. 35). Pearlite, because of its hard cementite plates, interferes with

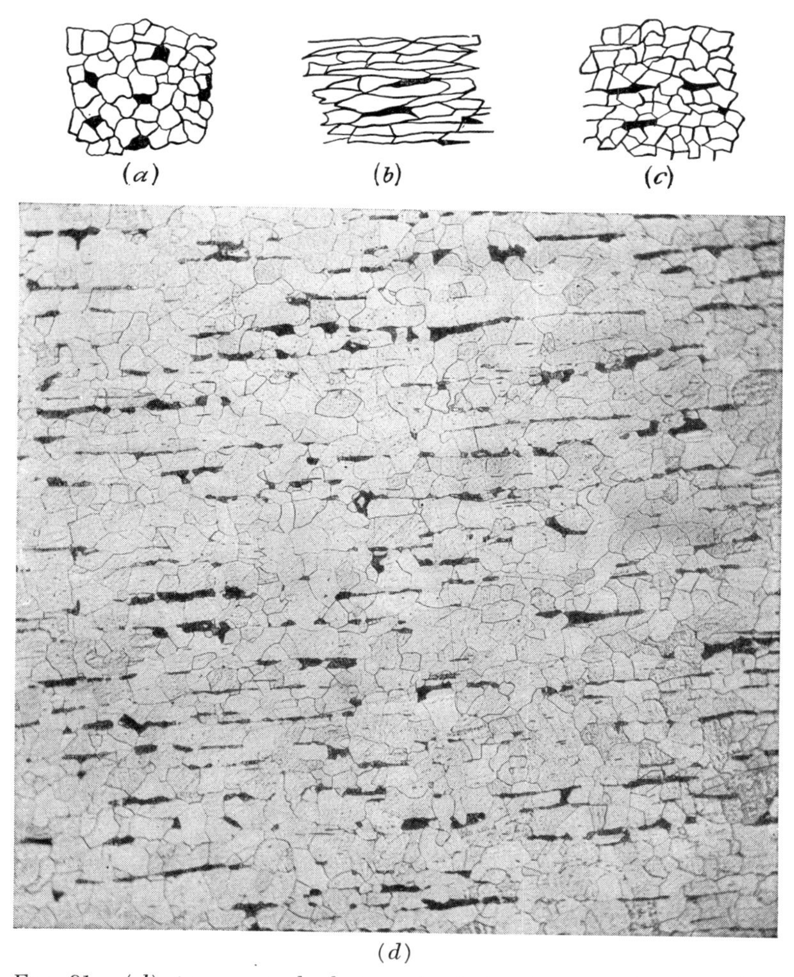

Fig. 81. (*d*) is an actual photograph of the structure sketch in (*c*). (*Courtesy General Electric Company.*)

drawing. Therefore, a different mode of heat-treatment is necessary, which is resorted to before drawing and between the drawing operations. This consists of heating to a quite high temperature (1500°F. to 1950°F.), which puts the carbon in solution; this is followed by a light quench in air or molten lead, conditions being adjusted just right to obtain a structure resembling sorbite. As outlined in Chap. 7, very careful control is necessary. This process is known as *patenting*—sometimes referred to as *sorbitizing*.

A glance at the structure of sorbite (see Fig. 43) shows that this would be an ideal condition for drawing, there being no large plates or particles of cementite to interfere. This grade of wire is usually finished in the drawn condition (not annealed after final drawing), and its physical properties have been changed from a tensile strength of 95,000 p.s.i. and an elongation of 10 per cent to 210,000 p.s.i. and an elongation of 2 per cent.

Uniformity of Grains

Looking at the above metal in the light of grain structure, it would be well to explain that every grain of the "sorbitized" metal is of identical material. While it was not explained in the text, the photomicrographs (see Figs. 34 and 43) show that rate of cooling has a great deal of influence on the structure of the individual grains.

If a plain-carbon steel of more than 0.20 per cent carbon is cooled slowly through the critical range, it will consist of separate grains of ferrite and coarse pearlite. If it is cooled at a rate that will cause the formation of very fine pearlite, in most cases every grain will consist of the same material—very fine pearlite—and this results in much greater strength (the weaker pure ferrite grains being absent). This explains the strengthening effect of quenching and of normalizing, and the similar effect of casting against a "chill."

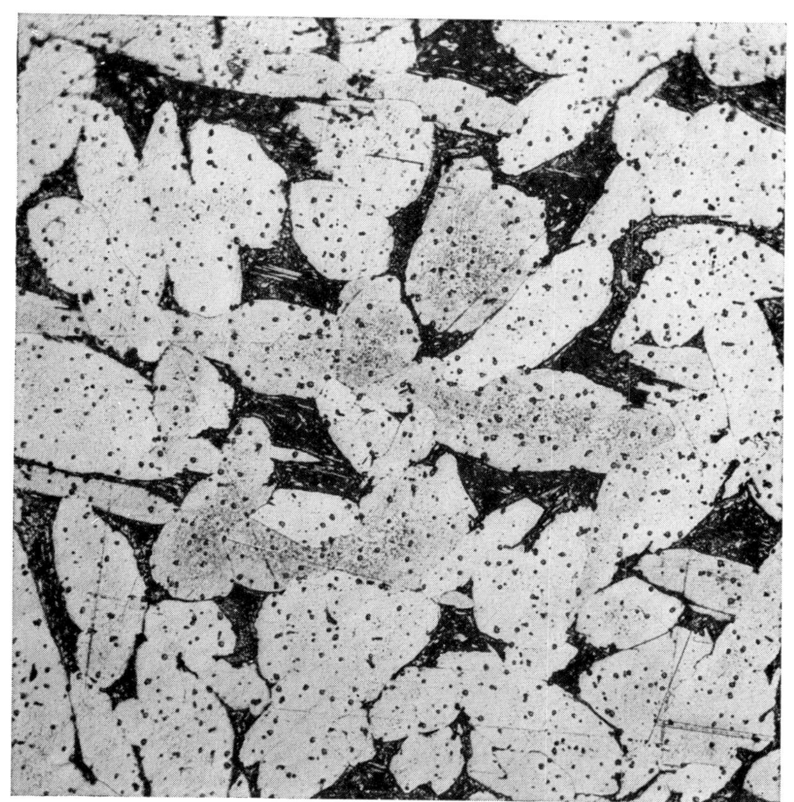

FIG. 82. Aluminum bronze at 150 diameters, furnace-cooled from 1815°F. White crystals are α solution. Dark crystals are β solution. Brinell hardness, 130. (*Courtesy General Electric Company.*)

Chills

Chills are pieces of iron coated with a clay wash so as to make them nonadherent, placed in the casting surface of sand molds either to improve the quality of the casting at that point or to prevent cracking of the casting. For instance, the old-fashioned cast-iron, railroad-car wheels were cast against a chill

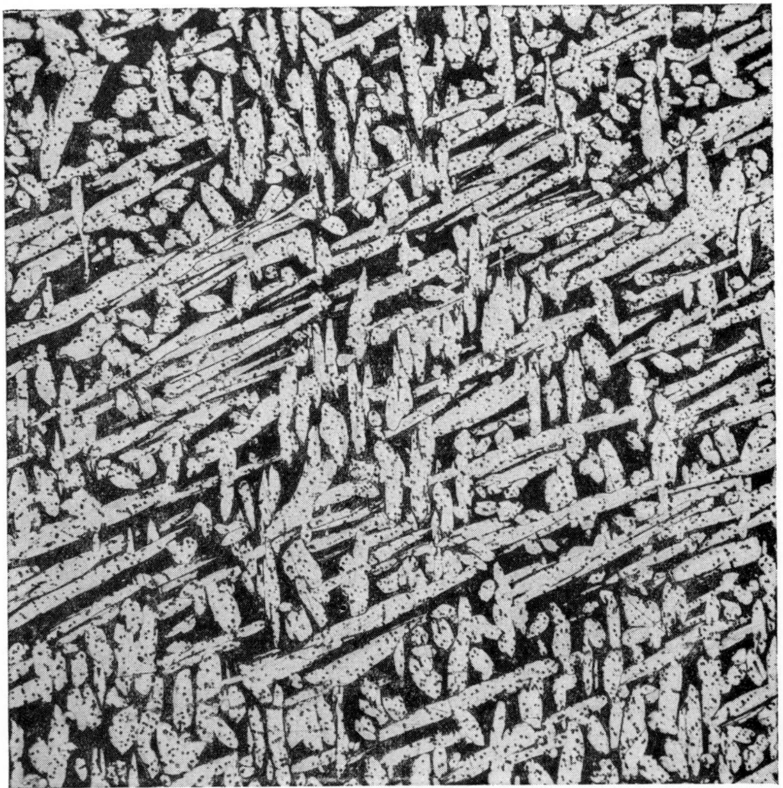

FIG. 83. Aluminum bronze at 150 diameters air-cooled from 1815°F. Crystals are much smaller. (*Courtesy General Electric Company.*)

on their circumference, the result being a resilient interior and a hard-wearing surface on the outer edge.

Permanent Molds

Many alloys are cast in *permanent molds,* molds made of tungsten steel, in order to get the strengthening effect of quick cooling (as well as closer tolerances in dimension). Aluminum bronze cast in permanent molds is made from a different mixture

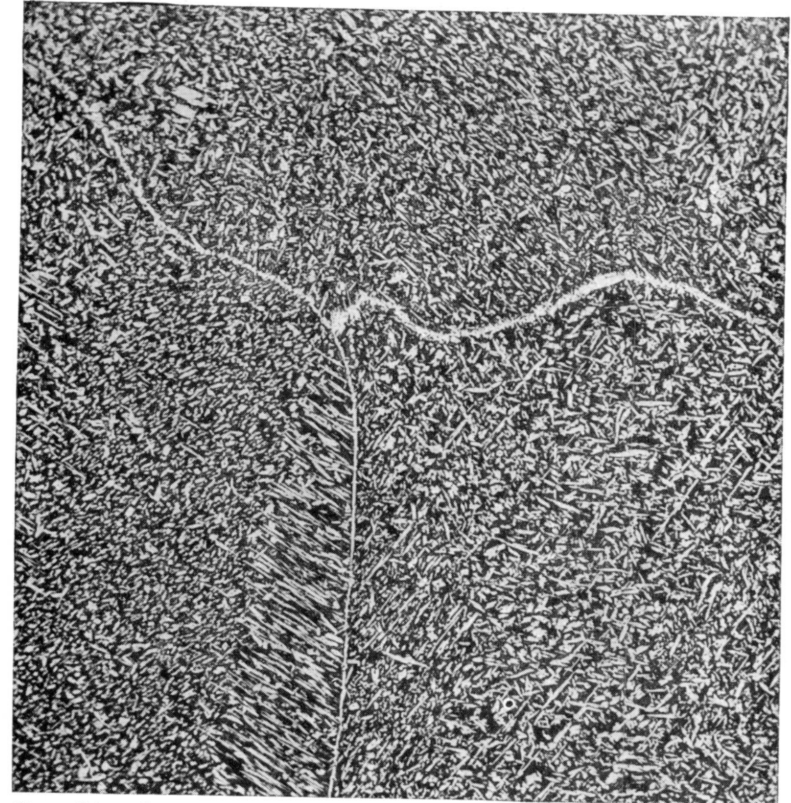

FIG. 84. Aluminum bronze at 150 diameters, water-quenched from 1790°F. (a draw at 1200°F. did not change the hardness—Brinell hardness, 166). (*Courtesy General Electric Company.*)

than that for sand cast, the difference being the addition of a "hardener" in the latter case. The effect of chilling on aluminum bronze is well shown by the photomicrographs (Figs. 82, 83, and 84). Figure 82 shows the alloy as slowly cooled and consisting of alpha solution (dark) and beta solution (white)[1] each

[1] See p. 60 for explanation of α, β, etc., solutions.

as separate crystals. Upon faster cooling, air-cooled (see Fig. 83), the structure has become uniform but is still rather coarse. Finally, Fig. 84 shows the fine structure caused by a water quench. All three of these heat-treatments were made on the same batch of metal.

The importance of *grain size* has become increasingly evident since investigations were started several years ago by McQuaid

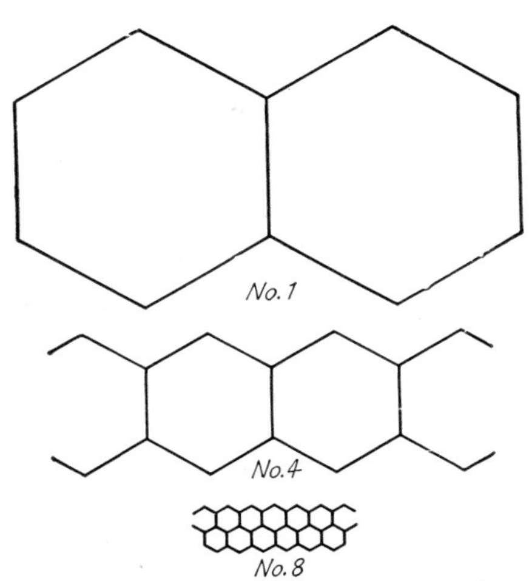

Fig. 85. Grain sizes. Austenitic grain sizes are usually given according to Timken, ASTM index, of which this sketch shows sizes 1, 4, and 8 at 100X magnification.

and Ehn[1] to determine the original grain size and its influence on steel. Figure 85 shows a few of the eight different grain sizes that have become standard. They are based on a magnification of 100 diameters. As explained under *annealing* and *hot-working*, grain growth takes place above the transformation temperature. However, in some steels this does not begin until

[1] H. W. McQuaid and E. W. Ehn, Effect of Quality of Steel on Case Carburizing Results, *Trans. Am. Inst. Mining and Met. Engrs*, Vol. 67, 1922, p. 341.

about 500°F. above that temperature. While it is true that fine-grained steels are stronger, the coarse-grained do have a few advantages. They machine well (because of more granular nature); they harden deeper; and carburizing penetrates deeper in them.

In regard to *grain shape,* it would be well to call attention to the fibrous nature of the grains in a forging or rolled shape. Notice this in Figs. 66, 81, and 125; also note a similarity in the structure of wrought iron. Rolled shapes and forgings have directional strength, greater strength in the direction of flow than

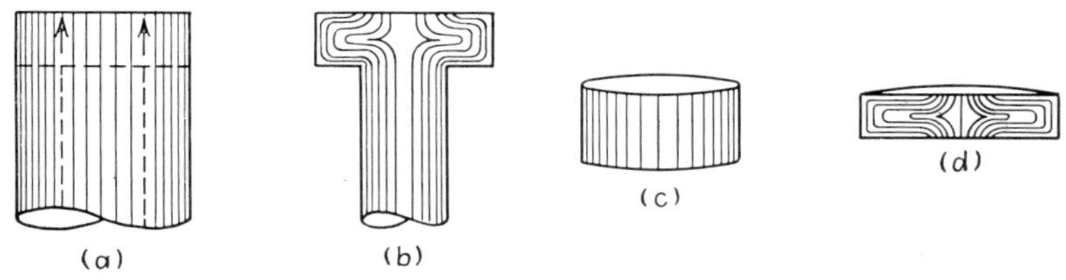

FIG. 86. Sketch illustrating directional properties in forgings.

transversely to that direction. This fact is always taken into account in planning their manufacture. The metal should be caused to flow in the direction which will meet the greatest stress. In the case of wire the desired direction is very obvious. The same principle must be taken into account in the design and production of all forgings (see Fig. 86).

The sketch at the left (see Fig. 86*a*) shows the fibers in an ordinary rolled bar. If now the metal indicated by the dotted line were removed by machining to make a bolt, it can be seen that no advantage has been taken of the directional strength as indicated by the fibers. In fact, the bolt's greatest stress (on the head) comes in line with the fibers, and the head could be sheared off much more easily than if the head were put on by upset forging. In that case the fibers in the head would be spread outward as in (*b*) and resist such shearing action. The same principle holds good in the forged gear blank at the right. The teeth will be much stronger when cut in the forging (*d*) than in a disk sawed from a rolled bar (*c*). Note that the center of the gear wheel may not be as strong when forged, but that doesn't matter, since it will be bored anyway.

DEFINITIONS

The word **crystal** can have two meanings: In the broader sense it means any portion of a metal or other solid in which the atomic arrangement is regular and natural; the narrower meaning describes the geometric solid that forms under favorable conditions of free space, time, and surroundings and that is bounded by plane surfaces at definite angles to each other, characteristic of the substance.

A **grain** is an individual crystal of metal, the word *crystal* being used in its broadest sense.

By **space-lattice** is meant a unit pattern of atomic spacing and arrangement. It suggests a rigid framework in space, on which atoms "hang."

Patenting.[1] A heat-treatment applied to medium- or high-carbon steel before the drawing of wire or between drafts. It consists in heating to a temperature above the transformation range, and then cooling to a temperature above that range in air or molten lead or salts maintained at a temperature appropriate to the carbon content of the steel and to the properties required of the product.

Process Annealing.[1] In the sheet and wire industries. A process by which a ferrous alloy is heated to a temperature close to but below the lower limit of the transformation range and is

[1] "Metals Handbook," pp. 10–11, American Society for Metals, Cleveland, 1948.

subsequently cooled. This process is applied in order to soften the alloy for further cold-working.

In heating processes of cold-worked material:[1]

Recovery is a change in properties, particularly those determined by the state of internal strain, of the cold-worked steel at low annealing temperatures, unaccompanied by any detectable change in the microstructure of the cold-worked metal.

Recrystallization is the development of an entirely new grain structure by heating to a higher temperature than that required for recovery.

Grain Growth of steel after recrystallization is nothing more than continued growth of the recrystallized grains.

The temperature at which work ceases is known as the *finishing* temperature.

QUESTIONS

1. Explain why fine-grained steel is stronger than coarse-grained steel.
2. Why are grain boundaries the weakest part of the metal at high temperatures?
3. Name a few points in favor of coarse-grain structure.
4. What is the purpose of chills?
5. Name two advantages of casting in permanent molds as compared with casting in ordinary sand molds.

REFERENCES

GROSSMAN, M. A., "Principles of Heat Treatment," American Society for Metals, Cleveland, 1940.

SEITZ, FREDERIC, "The Physics of Metals," McGraw-Hill, New York, 1943.

TEICHERT, E. J., "Ferrous Metallurgy," 2d ed., vol. III, McGraw-Hill, New York, 1944.

WILLIAMS, R. S., and V. O. HOMERBERG, "Principles of Metallography," McGraw-Hill, New York, 1948.

[1] E. J. TEICHERT, "Ferrous Metallurgy," 2d ed., New York, McGraw-Hill, pp. 190–191, 1944.

Chapter 10

ALLOY STEELS

So far, the discussion has covered only carbon steels, *i.e.*, steels in which carbon is the only influential element in the iron—manganese and silicon being present as deoxidizers principally, and all other elements, if present, being considered impurities. In this chapter will be briefly outlined the effect of several elements that have such a useful influence when added to steel that few industries could get along without them. They are all on the priority list in wartime. According to effects, alloys may be divided into:

1. Those which strengthen ferrite—nickel, manganese, silicon, chromium, aluminum, and cobalt.

2. Those which form carbides—molybdenum, chromium, vanadium, tungsten, and manganese.

3. Those which have a beneficial influence as *scavengers*, in removing oxides—aluminum, vanadium, and silicon.

4. *Graphitizers*—those which tend to drive carbon out of the combined form—nickel and silicon.

Most alloys are used in the heat-treated condition, and therefore their influence on the iron-carbon diagram will be studied first. As nickel furnishes a typical case and the least complicated, it will be the first alloy considered.

NICKEL

Nickel lowers the critical range, the point of transformation of gamma to alpha iron, in proportion to the percentage of it present. At the same time the eutectoid point is shoved to the left. This is illustrated in Fig. 87, which shows the change pro-

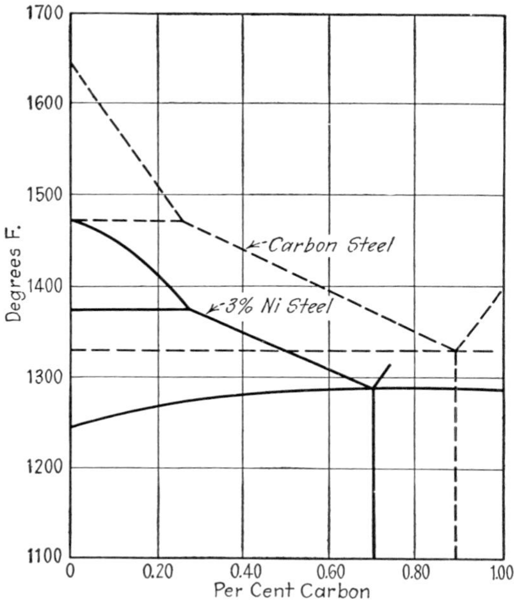

FIG. 87. Comparative location of critical changes on heating 3 per cent nickel steel and straight carbon steel. (*From Bullens, "Steel and Its Heat Treatment," 2d ed., Wiley.*)

duced by 3 per cent nickel. One advantage of the addition of nickel is apparent at once. Heat-treating can be done at lower temperatures, thus avoiding some of the warping and danger of cracking. The lower carbon content also removes some of the danger of cracking.

Not only does nickel lower the A_1 line, it also greatly increases the difference between the *Ac* and *Ar* lines. It increases the *lag*, as it is called (which, as was learned, was a matter of not much more than 20° in carbon steels). For example, if a 10 per cent nickel steel is cooled from a temperature above the critical range, magnetism will not be apparent until about 752°F. With the A_2 temperature for this steel at about 1292°F., this indicates a lag of about 540°. The lag increases with increasing nickel content until at about 25 per cent nickel the steel is not influenced by a magnet, even at 0°C. This means that the paramagnetic austenite (gamma iron) has been brought down to room temperature, which in plain-carbon steels cannot exist below 1333°F. except for a matter of seconds, as indicated to the left of the line showing the beginning of austenite transformation on the S curve.

The following diagram indicates the effect of varying per-

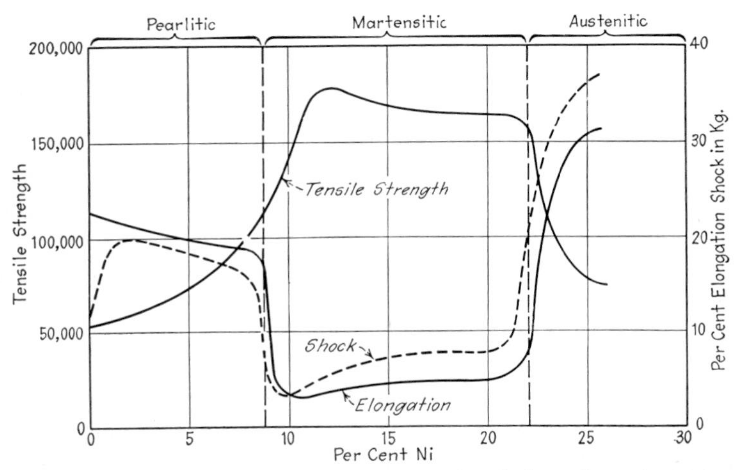

FIG. 88. Comparative physical properties of nickel steels with 0.25 per cent carbon. (*From Bullens, "Steel and Its Heat Treatment," 2d ed., Wiley.*)

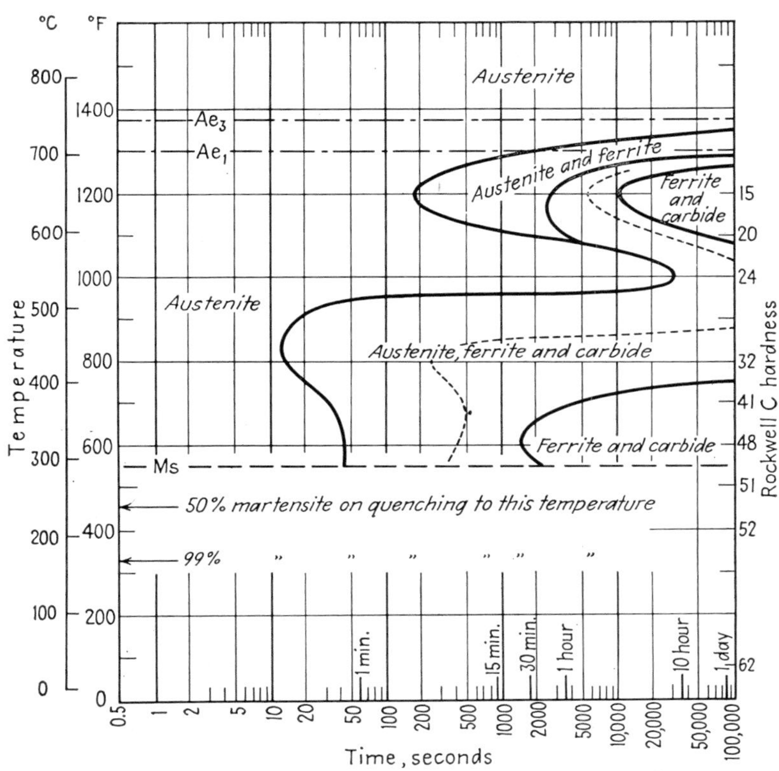

FIG. 89. Isothermal Transformation Diagram of a 4340 Steel. 0.42 per cent carbon, 0.78 per cent manganese, 1.79 per cent nickel, 0.80 per cent chromium, 0.33 per cent molybdenum. Grain size 7–8. Austenitized at 1550°F. (*Courtesy Dr. E. C. Bain and Carnegie-Illinois Steel Corporation.*)

centages of nickel and, to some extent, the properties of austenite. The diagram is drawn for a 0.25 per cent carbon content, not heat-treated. Notice how the tensile strength rises graually with increasing nickel content, reaching a maximum of 180,000 p.s.i., with 12 per cent nickel, and with decreasing elongation, all characteristic of martensite. When it is considered that this martensite is obtained on slow cooling, and when it is also considered that with increasing the nickel content to 22.5 per cent the austenite does not transform at all, it is obvious that alloys (nickel in this case) can greatly distort the S curve as compared with the curves for plain-carbon steel. Such distortion is illustrated by Fig. 89, which is a T.-T.-T. diagram for a nickel-chromium-molybdenum steel. Among other features, notice that there is a possible delay of almost 200 sec. in "getting by the knee of the S curve."

Properties of Austenite

Figure 88 is useful in showing the properties of austenite. At first glance it would seem to resemble pearlite, its tensile strength being just a little higher and its ductility (elongation) and impact resistance considerably higher. However, the fact that it is so drastically affected by cold-work causes it to have altogether different applications. It cannot be hardened by quenching as long as it remains austenite. It can be hardened only by cold-working. It cannot be lifted by a magnet, but it is not entirely nonmagnetic, since it will allow passage of magnetic lines of force. To sum up, austenite is quite a different metal from the iron we are accustomed to—almost resembling copper more than our common (alpha) iron.

There are two reasons for the beneficial effects of nickel on the properties of steel:

1. Nickel and iron are soluble in each other in all proportions, and nickel strengthens ferrite.

2. The lowering of the critical range and the moving of the eutectoid point to the left allows the formation of pearlite with less carbon, and the moving of the S curves to the right allows martensite to form with a much less drastic quench—or with no more quench than cooling in air if the nickel content is high enough.

Heat-treatment of Nickel Steels

Thus it is evident that nickel is a valuable alloy, even without heat-treatment. As to heat-treatment of nickel steels, the first conclusion would be that they can be quenched from a lower temperature, and that is substantially true. One modification enters here. Nickel steels require much more time for transformation; in order to save time, they are often heated well above their theoretical critical point, then cooled to a temperature just above their A_r point, and quenched. Tempering and toughening are carried on exactly as with carbon steels.

A chart is given here (see Fig. 90) showing the effect of quenching and tempering operations on nickel steel. The advantages of using nickel can be studied by comparing these values with corresponding values in a plain-carbon steel of almost the same carbon content (see Fig. 44).

Nickel steels lend themselves very well to case carburizing and nitriding because the core is not damaged by high temperatures. This property is particularly applicable in the case of gears, the surfaces of which can thus be made extremely hard and wear-resistant. The nickel in this instance is 3 to 4 per cent.[1] In fact, most of the nickel steel used has a low percentage of nickel, corresponding to the pearlitic range of Fig. 88. Steels in the martensitic range are seldom used because they are so difficult to work or machine. In the austenitic range, however, there are several very special steels. Steels of 25 per cent nickel and 0.7 to 0.9 per cent carbon have great resistance to oxidation at high temperatures and are used for such purposes as automobile valve stems. An alloy of 36 per cent nickel, known as *Invar*, has a negligible coefficient of expansion

[1] Some products of 3½ per cent nickel steels are highly stressed bolts; heat-treated tubing subjected to torsional stresses, such as automobile front axles and propeller-shaft tubes, with carbon 0.35 to 0.40; quenched and tempered shafting; connecting rods; and universal-joint yokes.

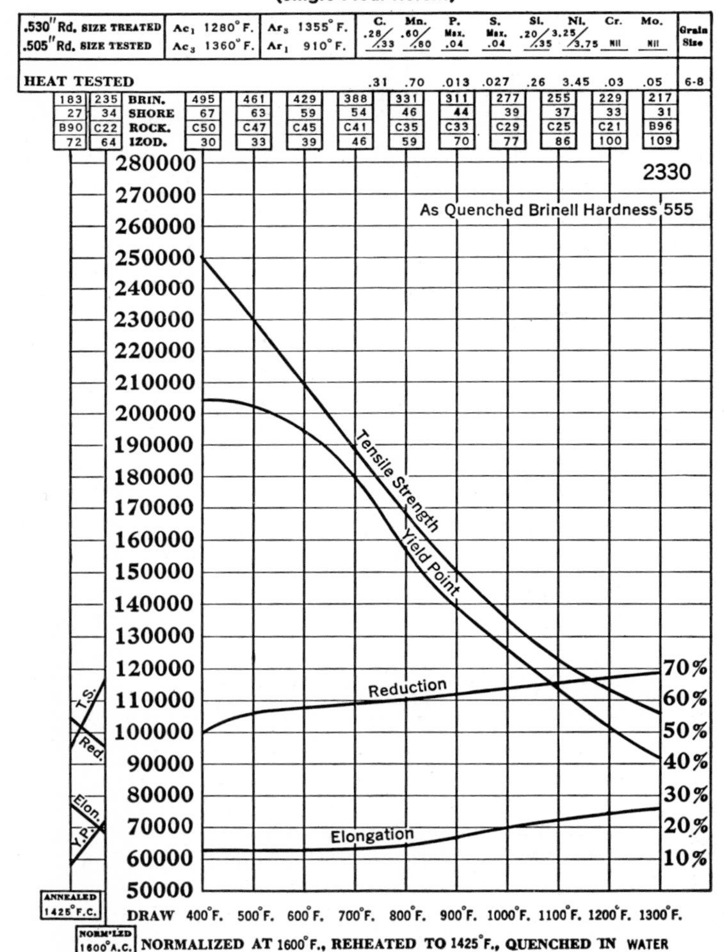

FIG. 90. Properties Chart. (*Reprinted from "Modern Steels and Their Properties," Handbook 268, Bethlehem Steel Company, 1949.*)

at room temperature, which makes it valuable for measuring instruments such as clock pendulums, balance wheels of watches, etc. Another alloy, *platinite,* has the same expansion rate as glass and is therefore used for lead-in wires in electric lamps. Perhaps the most extensive use of nickel is in conjunction with other alloys, such as chromium. These alloys will be described after the discussion of chromium.

CHROMIUM

Chromium, like nickel, is one of the key metals essential to the development of industry. It may be used either alone, in which case it profoundly modifies the properties of ordinary carbon steel, or in conjunction with nickel, tungsten, vanadium, molybdenum, etc. It has a greater variety of applications than nickel because of its corrosion resistance and also because of its ability to resist the effects of heat. Both nickel and chromium are extensively used in plating; as would be expected, the latter is harder and much more resistant to corrosion.

As to the influence of chromium on the iron-carbon diagram, Bullens says, "Chromium in steel has the characteristic function of opposing both the disintegration and reconstitution of cementite."[1] It resembles nickel in lowering the *Ar*, but it also has the opposite effect of raising the Ac_1.

The constitution diagram for the iron-carbon-chromium system is more complicated than for the Fe-C-Ni system because chromium forms carbides. One (Cr_3C) is similar to Fe_3C, but in the presence of iron can be Fe_2CrC or Cr_2FeC and is written $(Fe,Cr)_3C$. There is another complex carbide of this composition, $(Cr, Fe)_7C_3$. These and the fact that there is a limit to the solubility of chromium in gamma iron serve to

[1] D. K. Bullens, "Steel and Its Heat Treatment," Wiley, New York, 1918.

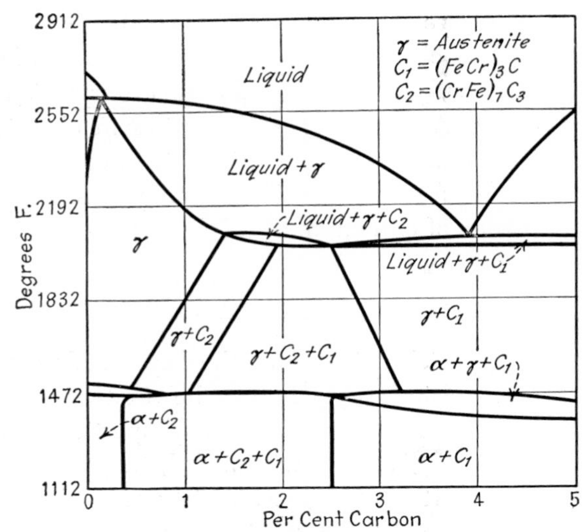

FIG. 91. Section of Fe-Cr-C system at 5 per cent chromium. (*"Metals Handbook," American Society for Metals,* 1939).

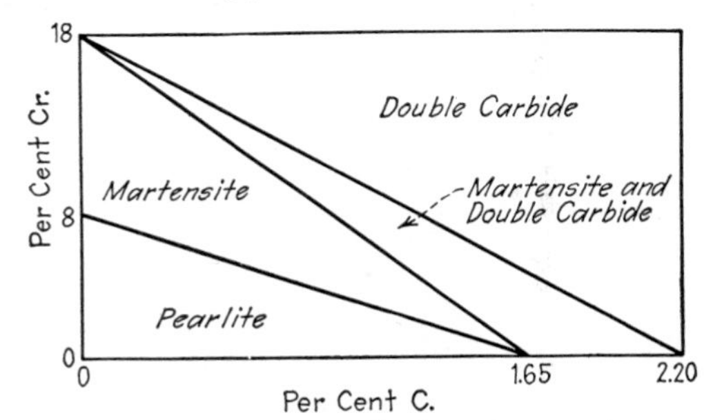

FIG. 92. Microscopic constituents of chromium carbon steels. (*From Bullens, "Steel and Its Heat Treatment," 2d ed., Wiley.*)

complicate the diagram. Figure 91 shows a small portion of the diagram for 5 per cent chromium with varying carbon content. Figure 92, taken from Bullens' work on heat-treating, is useful in showing the variety of structures possible with chromium steels.

Effect of Chromium Variations

Steels with a chromium content up to 0.50 per cent are used where a moderate increase in hardenability is desired, along with an increase in toughness. It tends to reduce the grain size,[1] and it might be mentioned here that chromium makes steels *deep hardening*, meaning that they possess a much greater penetration of hardness from the surface inward than would be possible with straight carbon steels.

Chromium carbide imparts great wear resistance. The teeth of a steam shovel are a good example. Steels for such uses are in the range of 1 to 2 per cent chromium. One of the most indispensable uses of chromium as an alloy is for ball and roller bearings; they contain about 1 per cent carbon and 1½ per cent chromium (the bearings being a little higher in both elements than the balls). Along this line should be mentioned chisels, files, high-speed drills, rolls, dies, etc. The next higher range includes high-speed cutting tools. These all contain about 4 per cent chromium but always in conjunction with some tungsten and molybdenum, and so they will be mentioned later. Stainless steels also will be mentioned later, in connection with their respective additional elements.

The word *stainless* suggests resistance to corrosion. Increasing the chormium content between 4 and 14 per cent rapidly increases the resistance to acids (even the stronger acids, such as nitric), and such steels find a wide use in oil-refining stills and in many chemical processes, and their use in cutlery and similar products is rapidly increasing.

Similar to these are the heat-resistant chromium steels (duraloy, fabrite, etc.), containing 22 to 28 per cent chromium and

[1] It is a very valuable alloy for reducing the size of carbon flakes in gray cast iron (see Figs. 97 and 98). Besides, high-chromium cast iron is one of the most abrasion resistant of all ferrous metals. It is used in sandblast equipment.

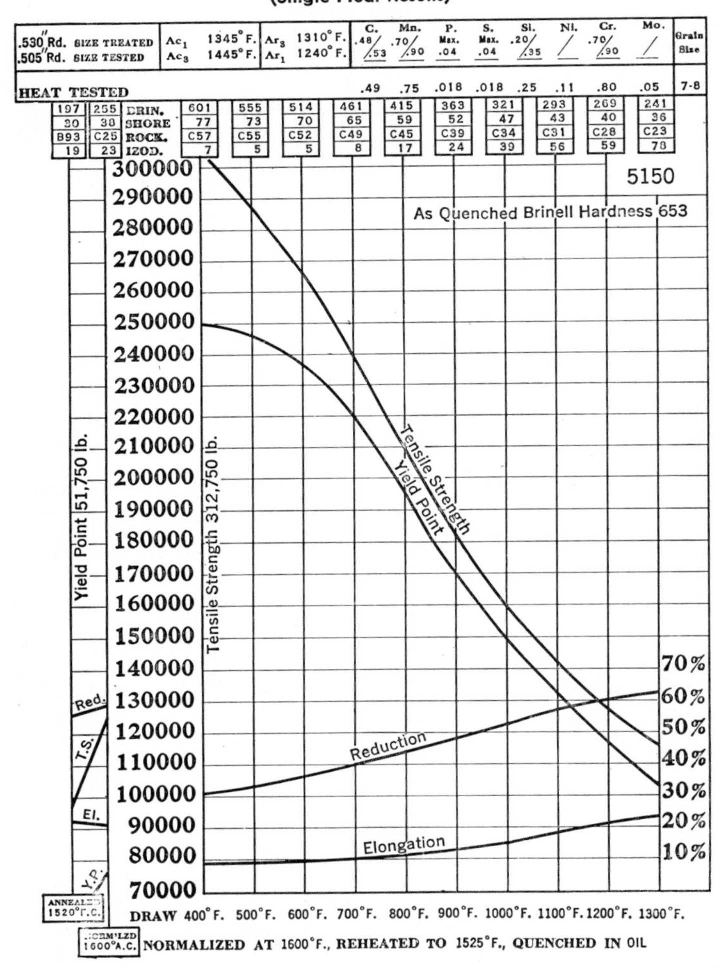

FIG. 93. Properties Chart. (*Reprinted from "Modern Steels and Their Properties," Handbook 268, Bethlehem Steel Company,* 1949.)

a variable content of carbon from 0.2 to 1 per cent (the lower percentages of carbon are used in pieces that have to be machined and the higher in those where excessive hardness is not objectionable). While the mechanical properties of these very-high-chromium steels apparently are no better than those of corresponding straight carbon steels, they have an extremely great resistance to oxidation at high temperatures and may be repeatedly heated for very long periods at temperatures as high as 1800°F. and in some cases 2000°F. (A recent development contains 37 per cent Cr and 7 per cent Al. It is used for electric-furnace resistors and resists oxidation up to 2370°F.)

CHROMIUM-NICKEL STEELS

Bullens says:

> Chromium-nickel steels probably represent the best all-around alloy steels in commercial use for general purposes. Chromium-nickel steels of suitable composition appear to have the beneficial effects of both the chromium and nickel, without the disadvantages which are inherent in the use of either one separately.[1]

In many ways the two elements are opposites. Nickel forms no carbides; its influence is felt only in ferrite and in the way it distorts the iron-carbon diagram (see Fig. 87). On the other hand, the principal influence of chromium is in forming complex carbides. It does, however, have some beneficial influence on ferrite. Probably the best ratio is 2½ parts nickel to about 1 part chromium.

Effect of Chromium and Nickel Variations

There are several applications of chromium-nickel steels. In the SAE 3100[2] series, 1.25 per cent Ni, 0.6 per cent Cr, the steels

[1] Bullens, *op. cit.*, p. 349.

[2] See Appendix for the significance of these SAE figures.

with about 0.30 per cent carbon are used for axles, drive shafts, stresser bolts, studs and nuts, automobile steering knuckles and arms, drill collars in oil-well-drilling equipment, etc. When annealed, they can be cold-pressed into parts such as automobile frames, which are subsequently quenched and tempered.

Steels still in this Cr-Ni range, but with slightly higher carbon (0.40 per cent), are well suited for aircraft, bus- and truck-engine crankshafts, power-shovel parts, etc. When tempered to about Bhn 400, they show excellent wear resistance in excavating- and road-building-machinery parts. This grade of steel is also used for automobile intake valves.

In a higher range, 2 per cent Ni and 1 per cent Cr (SAE 3200 series), usually with a 0.40 to 0.50 per cent carbon, some of the most dependable of all the chromium-nickel steels in respect to strength and toughness may be obtained. They are used for automobile drive shafts, connecting rods of aircraft radial engines, transmission gears, etc. Still higher, 3½ Ni, 1½ Cr, are steels much more difficult to handle, but they find much use in connecting rods, rocker arms, propeller-hub cones, starter gears, etc. The effect of chromium on tensile strength and hardness, and the effect of nickel on ductility also the combined effect of the two elements, can be studied in Figs. 90, 93, and 94 (a small allowance should be made for difference in carbon content).

Perhaps the most interesting, metallurgically speaking, of the chromium-nickel steels are the stainless steels, containing about 18 per cent Cr and 8 per cent Ni (often termed 18-8). Besides their resistance to corrosion, they are of great interest in this course because of the way in which they are hardened. These are austenitic steels and therefore cannot be hardened by quenching (this hardening depending on the transformation of austenite, as has been discussed previously). Such stainless steels are hardened by cold-working. However, they also have

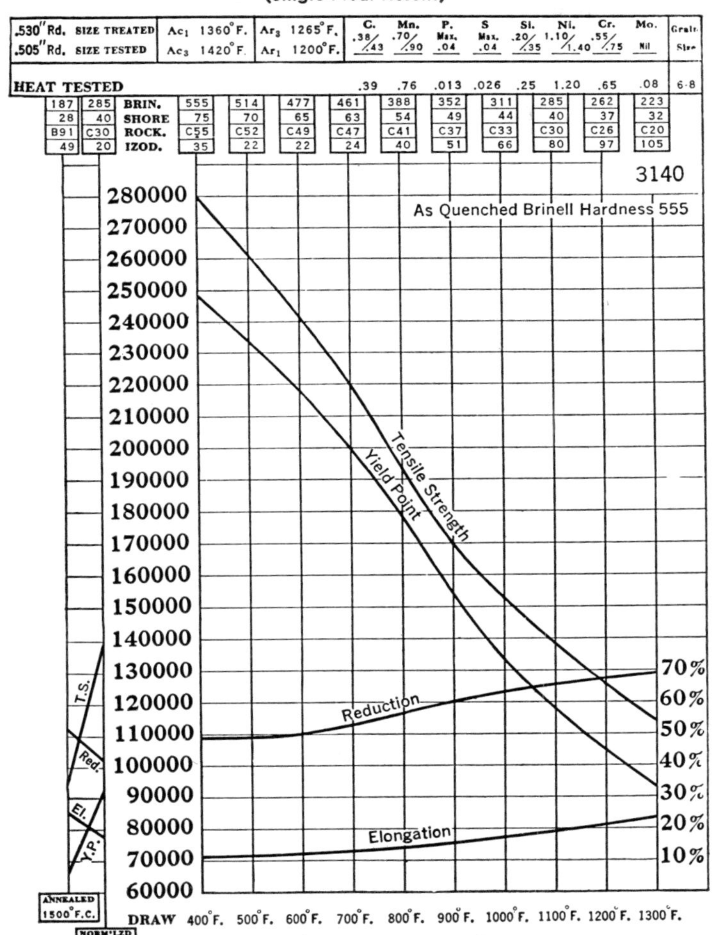

FIG. 94. Properties Chart. (*Reprinted from "Modern Steels and Their Properties," Handbook 268, Bethlehem Steel Company, 1949.*)

some hardening quality in addition to the strain imposed by grain deformation.

Austenite is unstable, considerably more so than martensite. In the grain boundaries, therefore, where the distortion of atomic regularity is greatest, there is a chance for a small part of the austenite to transform into martensite during the *grinding* between grains. That is one reason that cold-rolling raises the elastic limit from 80,000 to 150,000 p.s.i. in the stainless steel used on streamlined trains.

In fact, this steel work hardens so quickly that center punching for a drill hole must be done very lightly in order to start the drill at all, for the punch mark will harden that spot. Drilling itself is a very difficult operation, because there is a small area between the flutes of the drill that is not cutting. This does not make so much difference with a large drill as with a small one. The same idea applies to machining—the tool must always be cutting, never rubbing.

VANADIUM

This element is one of the more expensive of the alloys mentioned in this text, but its usefulness is attested by the great demand for it. It is seldom used in percentages greater than 0.20 per cent (except in high-speed-tool steel). Perhaps its most marked influence is in preventing large grain growth. This is especially true when it is used in combination with chromium; and therefore chrome-vanadium steels are well adapted to uses where a casehardened surface and strong core are required, in automobile gears, for instance. The advantage is that the high temperature necessary for carburizing does not coarsen the grain.

The tendency toward fine grain structure that vanadium imparts makes it very beneficial for use in steel castings, for it prevents such a condition as was shown in Fig. 61. Nickel is often used in conjuction with it for castings, a typical analysis being carbon, 0.28; Mn, 1.00; Ni, 1.50; and V, 0.10. After normalizing and tempering, a tensile strength of 93,850 p.s.i., 28 per cent elongation, and 50 ft.-lb. izod impact can be expected.[1]

Vanadium resembles chromium in regard to forming carbides that are hard to break up. It has the lowest eutectoid point of any of the common alloys. One per cent vanadium lowers the eutectoid point of the iron-carbon diagram to 0.65 per cent carbon. Uses for vanadium steels include locomotive forgings, automobile crankshafts, steering arms, connecting rods, axles, etc.

TUNGSTEN

Tungsten has been mentioned in connection with powder metallurgy and permanent molds. Everyone knows that electric-lamp filaments are tungsten, and so it is no strain on the imagination to realize that tungsten stands high temperature. By far the greatest amount of tungsten, however, is used as an alloy in high-speed-tool steel. In this respect it seems practically indispensable, although when Japan imperiled our greatest source of supply,[2] efforts were made to substitute *molybdenum.* "Moly" can take the place of tungsten to quite an extent in high-speed-tool steel; but certain difficulties are encountered that make it impossible for molybdenum to displace tungsten entirely. The effect of tungsten on steel can be visualized by considering (*a*) its very high melting point, (*b*) the fact that it raises the critical points, and (*c*) that it forms complex carbides with iron somewhat in the manner of chromium.

[1] JEROME STRAUSS and GEORGE L. NORRIS, "Metals Handbook," p. 598, American Society for Metals, 1939.

[2] China used to supply one-half our tungsten.

In fact, tungsten carbide is one of the hardest substances known and is used in cutting tools. Being too brittle to be used alone, fine particles are imbedded in a cobalt matrix. This is done by mixing the powdered carbide with powdered cobalt and sintering. In addition to being extremely hard, tungsten holds this hardness at high temperatures. At 1300°F. it retains a good cutting edge, and even at 1550°F. it retains as much hardness as ordinary high-speed steel does at 825°F. It also finds considerable use as a facing on dies for punching or forming metal.

"Secondary" Hardening

As was stated above, tungsten is quite an essential ingredient of high-speed cutting tools (see typical analyses under *molybdenum*). These are the tools that can cut metal fast enough so that they become red-hot. One thing to consider in the heat-treatment of such high-speed steel is that often when it is quenched, considerable austenite is entrapped, *i.e.*, not changed. Now, when the steel is heated again, either in tempering or in use, this austenite completes its transformation to martensite because the heat admits of atomic adjustment. (It is characteristic of many alloys to cause sluggishness in transformations.) Thus *the steel actually becomes harder on reheating.* This phenomenon is called *secondary hardening* and helps to explain why tungsten steel can be said to "get hot without losing its temper." It can not only get up to red heat, 1200°F., without losing its temper; but because of the peculiarity of *secondary hardening,* it can actually get harder in doing so. (See Appendix for heat-treatment of high-speed-tool steel.) High drawing temperatures are recommended to get rid of the retained austenite.

NOTE: This retained austenite can also change slowly at low temperatures. It expandes in so doing, and because the metal cannot "give," cracks are liable to result. The remedy is to draw soon after the quench.

Aside from its use as an alloy, tungsten is really an interesting metal in itself. It has the highest melting point of any of the metals (6098°F.) and so cannot be produced in any ordinary manner. (See *powder metallurgy* for its production.) The drawing of tungsten filament, red-hot through diamond dies, is really cold-working of the metal, when its high melting point is considered. It results in a fibrous structure very strong in the direction of drawing; in fact, tensile strengths of 590,000 to 650,000 p.s.i. have been reached.

MOLYBDENUM

Molybdenum is the only metal that can be substituted for tungsten in high-speed-tool steel, and it is very fortunate that the United States produces most of the world's supply. So far, a high-speed-tool steel has not been devised where all of the tungsten has been replaced by molybdenum, but the following analyses are typical, and it is possible in the near future that some combination can be discovered that does not need tungsten.

Former percentage		Recent percentage	
Tungsten	14–20	Tungsten	6–1.4
Chromium	3–4.5	Molybdenum	4–9
Vanadium	0.75–2.25	Chromium	4–3.8
Carbon	0.50–0.85	Vanadium	1–1.11
		Carbon	0.8–0.74

The molybdenum tool steels are not so easy to heat-treat; for instance, they must be heated in a reducing atmosphere or they will lose carbon. Tungsten steels do not require this

AISI-4640

(Oil Quenched)

PROPERTIES CHART

(Single Heat Results)

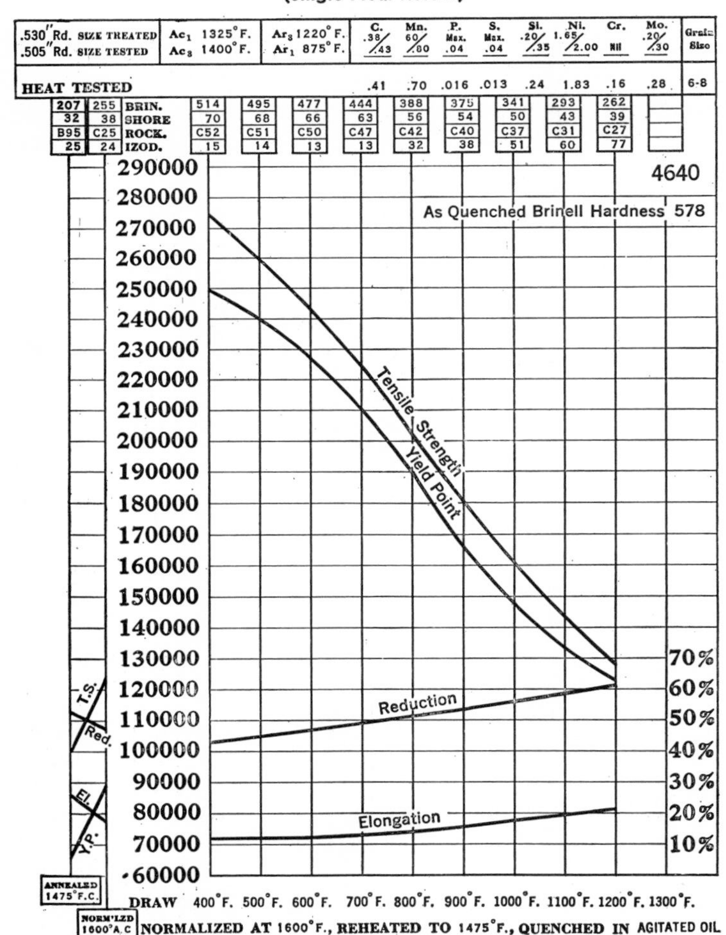

Fig. 95. Properties Chart. (*Reprinted from "Modern Steels and Their Properties," Handbook 268, Bethlehem Steel Company, 1949.*)

AISI-4340

(*Oil Quenched*)

PROPERTIES CHART

(Single Heat Results)

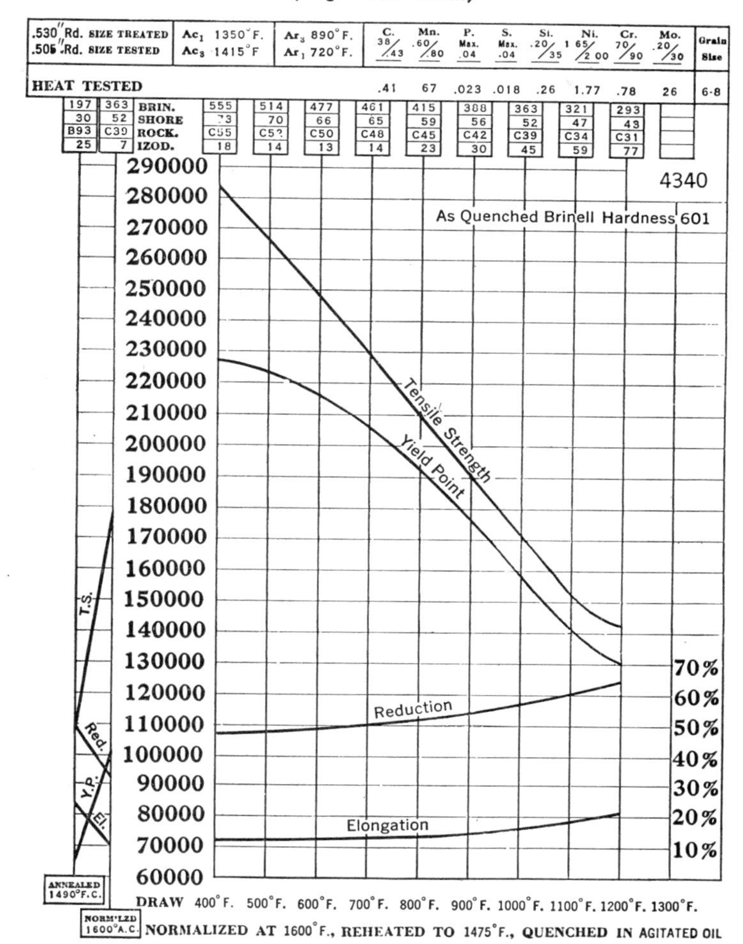

FIG. 96. Properties Chart. (*Reprint from "Modern Steels and Their Properties," Handbook 268, Bethlehem Steel Company,* 1949.)

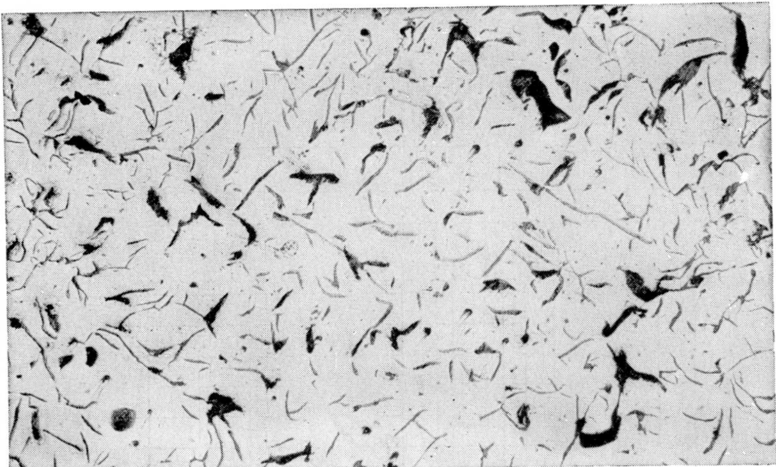

Fig. 97. Photomicrograph showing graphite distribution in unalloyed cast iron (unetched). (*From "Molybdenum in Cast Iron," Climax Molybdenum Company.*)

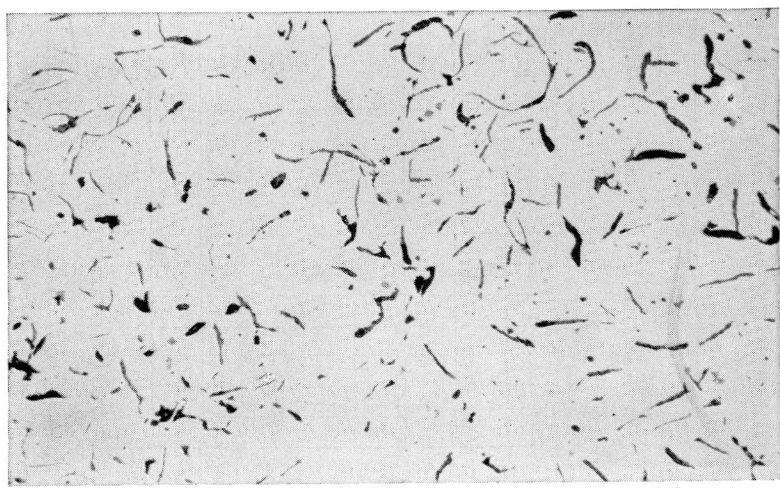

Fig. 98. Showing beneficial effect of 1.02 per cent molybdenum on graphite distribution. (*From "Molybdenum in Cast Iron," Climax Molybdenum Company.*)

care. Except for high-speed-tool steels, the use of molybdenum has been confined chiefly to steels of 1 per cent or less molybdenum content. It is practically always used in connection with other alloys, where it imparts a high elastic limit and tensile strength at higher temperatures without cutting down the elongation appreciably, as the preceding diagrams show (see Figs. 95 and 96).

Creep

An outstanding property of molybdenum steels is their ability to retain hardness and strength at high temperatures. In the manufacture of turbines this is a very valuable property because the efficiency of the turbine depends on its retaining all through its life the same tolerance with which it was built, regardless of high operating temperatures. Deformation due to stress and heat is known as *creep*. See definition, page 227. Molybdenum steels have about twice the creep strength of plain-carbon steels at temperatures above 900°F., and the usual moly content is around ½ to ¾ per cent.

Molybdenum, like chromium and nickel, has been used a great deal in alloying cast iron. The first two not only strengthen the ferrite, but, what is more important, cause smaller graphite flakes, making for more continuity in the iron matrix (see Figs. 97 and 98). Nickel also strengthens ferrite and has a beneficial effect on gray iron, but it has less effect on the graphite particles, on account of its being a graphitizer. Nickel and chromium together make a good combination. The former ensures machinability, which the latter might harm.

MANGANESE

Manganese was one of the first alloys to be used in steel. It has the same property that nickel has of lowering the critical

AISI-1340

(Oil Quenched)

PROPERTIES CHART

(Single Heat Results)

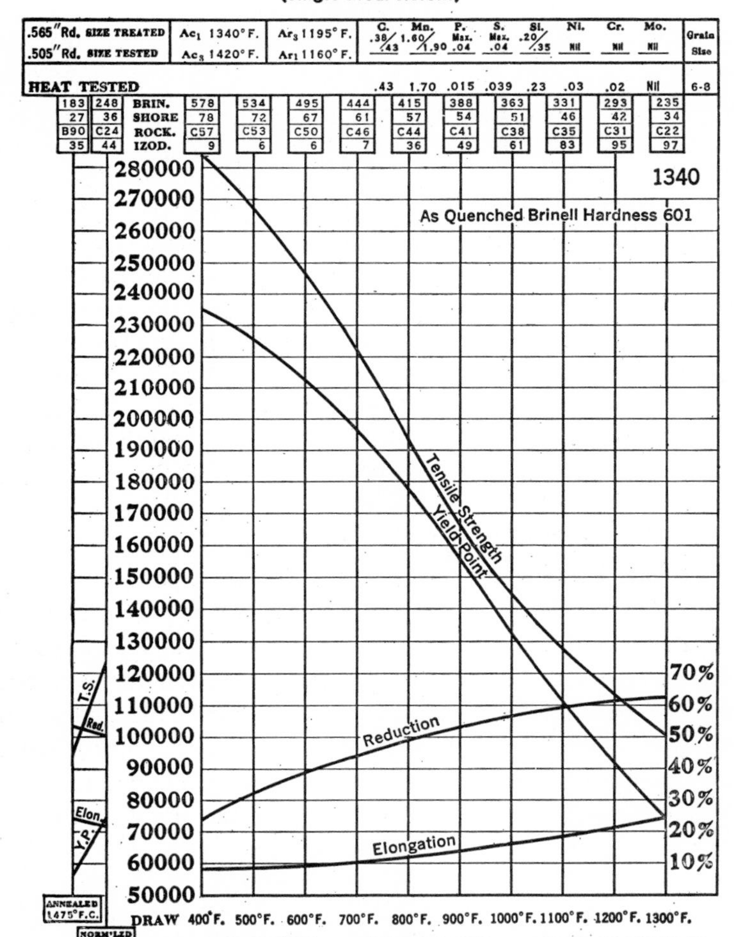

Fig. 99. Properties Chart. (*Reprinted from "Modern Steels and Their Properties," Handbook 268, Bethlehem Steel Company,* 1949.)

point to such a degree that with sufficient manganese (10 per cent) the steel is austenitic. Sometimes the use of nickel for this purpose is preferred because nickel does not form a carbide; but austenitic manganese steel is used extensively for railroad frogs and crossovers, rock-crusher parts, and power-shovel teeth; for it not only work-hardens quickly, but is resistant to wear. The more a manganese-steel railroad frog is pounded, the harder it gets, and such parts get a lot of pounding as the wheels go across.

The pearlitic (medium) manganese steels, with manganese content up to 2 per cent, seem sensitive in their heat-treatment, unless the ratio of carbon to manganese is carefully observed. However, they find a wide use and are obtainable in several grades. The chart in Fig. 99 describes one.

Manganese is used in small amounts in all steels as a deoxidizer and desulfurizer. Its requirements as such are governed by ratios to the amounts of sulfur and oxide present; the excess makes itself felt in the carbon ratio, and is thus considered as alloy.

SILICON

So far, except in the case of nickel, carbide-forming alloys have been treated. Silicon, as has already been seen (see page 40), has the opposite effect. It drives carbon out of combination with iron; and because of this effect, it has a very valuable application in electrical apparatus. A slight amount of combined carbon causes steel to retain magnetism. Inasmuch as alternating current magnetizes steel first in one polarity and then in the other, 60 times a second, any residual magnetism results in heat and is a waste of energy. This is known as *magnetic hysteresis.* Besides, silicon increases resistance to *electric* current flow, thus reducing the loss that results from induced eddy currents. There is still another advantage in

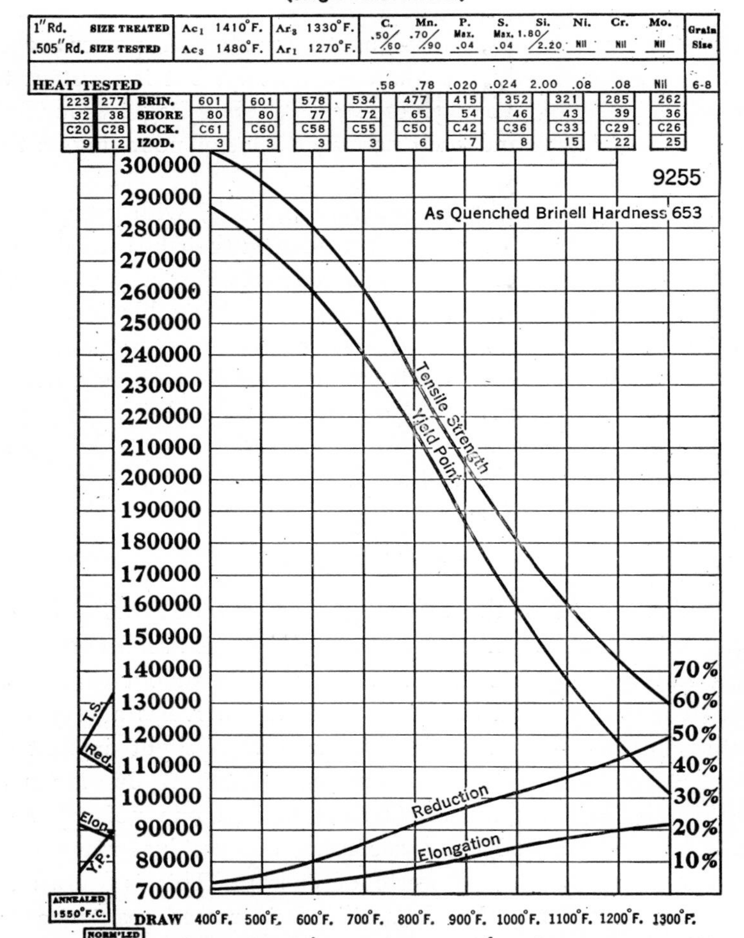

FIG. 100. Properties Chart. (*Reprinted from "Modern Steels and Their Properties," Handbook 268, Bethlehem Steel Company, 1949.*)

favor of silicon in magnetic circuits, and that is a possibility of obtaining large crystals, which increases magnetic permeability. Therefore, transformer laminations and armature punchings contain up to 3½ per cent silicon. This is a somewhat brittle but magnetically "soft" material.

Silicon also imparts another very valuable property, that of resistance to attack by acids. For this purpose it must be present in amounts from 12 to 17 per cent. Steels with this amount are always cast and find large application in chemical industries. Silicon strengthens ferrite and is used in small percentages, not over 2 per cent, in spring steels, usually in combination with vanadium. See Fig. 100 for properties of 2 per cent silicon steel.

Several other elements are used in alloying steels, but there is space only to mention them—cobalt, cerium, boron, and especially copper. The latter gives excellent corrosion resistance.

Depth-hardening

It was mentioned in Chap. 6 that uniform hardening of large sections was impossible in the case of plain-carbon steels. The rate of heat transfer prevents a uniform quenching rate throughout the piece. Here is where most alloy steels have a great advantage—not because the heat moves any faster, but because of their much slower rates of transformation.[1] This advantage is well illustrated by the comparison (see Fig. 101) of the hardening characteristics of two steels of practically the same carbon content, one plain carbon, the other containing 1 per cent chromium and 0.18 per cent vanadium, and both quenched in water from 1570°F.

It will be noticed that the surfaces of both steels show about

[1] Even the beginning of austenitic transformation is delayed. In the language of the S curve, the S's are not only shoved to the right, but they are also more widely separated.

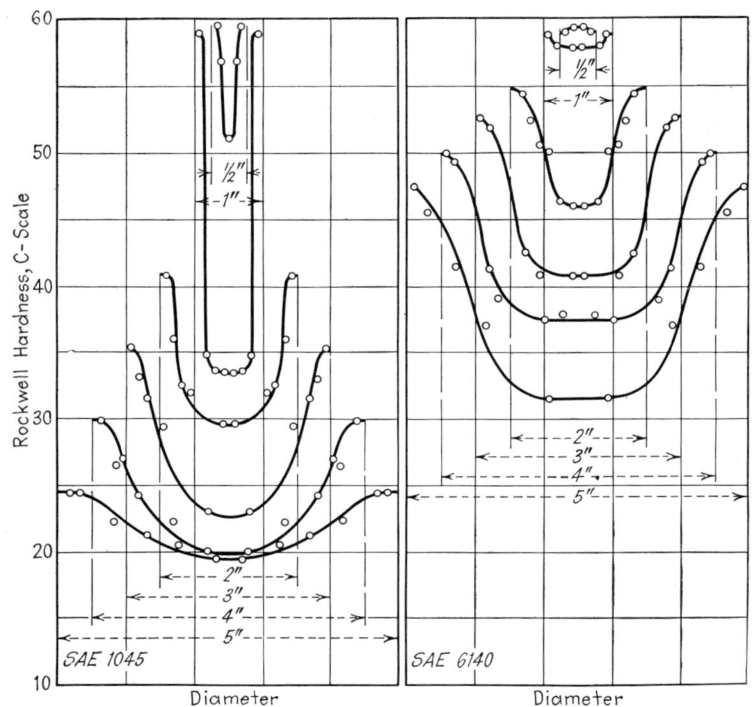

FIG. 101. Hardness surveys across diameters of bars of various sizes of 0.45 per cent plain-carbon steel (left) and chromium-vanadium steel with 0.40 per cent carbon (right) after quenching in water from 1575°F. (*By M. A. Grossman, from E. E. Thum, Editor, "Modern Steels," American Society for Metals.*)

the same hardness in the 1-in. and ½-in. sizes but that the carbon steel drops very markedly in hardness toward the center, while the alloy shows practically uniform hardness throughout. As to the larger sizes, it was noted on page 91 that on account of heat transfer, carbon steels of 2 in. or more in diameter cannot even be hardened satisfactorily at the surface, the heat flowing outward from the warm interior and slowing down the quenching rate. For this reason the *interior* of the 5-in. alloy piece shows a greater hardness than even the *exterior* of the 4½-in. plain-carbon piece.

Nickel and molybdenum also impart excellent depth-hardening properties.

QUESTIONS

1. In what respect are nickel and manganese alike in their effect on iron? in what respect unlike?
2. In what respect are silicon and nickel alike?
3. What is the composition of balls for ball bearings? of stainless steel? of high-speed-tool steel?
4. What is the highest tensile strength shown on a diagram in this chapter. What elongation accompanies it?
5. Manganese has three uses (deoxidizer, sulfur neutralizer, and plain alloy). Which is the most widely used?
6. Cite one or more allowing elements the effect of which is
   a. To increase hardenability ________________
   b. To strengthen ferrite ________________
   c. To form carbides ________________
   d. To reduce grain size ________________
   e. To deoxidize steel ________________
7. Molybdenum is said to promote air hardening of steel better than any other alloy. What advantage or disadvantage is this?
8. What is meant by secondary hardness?
9. What temperating temperature would you suggest to give an SAE 4340 steel a tensile strength of 200,000 p.s.i.? What yield point, elongation, and reduction of area would you expect?
10. Explain the advantages of alloys in producing uniform hardness when quenching large sections.

REFERENCES

ARCHER, R. S., J. Z. BRIGGS, and C. M. LOEB, JR., "Mobybdenum, Steels, Irons, Alloys," Climax Molybdenum Company, New York, 1948.

BAIN, EDGAR C., "Functions of the Alloying Elements in Steel," American Society for Metals, Cleveland, 1940.

PALMER, FRANK, R., "Tool Steel Simplified," The Carpenter Steel Company, Reading, Pa., 1937.

THUM, ERNEST, E., Editor, "Modern Steels," American Society for Metals, Cleveland, 1939.

Chapter 11
NONFERROUS ALLOYS

With the exception of copper and aluminum in connection with electric conductivity or iron in connection with magnetism, not very many metals are used commercially in their pure state. This is mainly because they are not strong enough. They can be strengthened by cold-working, but this is hardly practicable in the case of intricate shapes, nor is it generally desirable to use a metal in such a stressed condition that it is near the breaking point.

It was noted earlier that the strengthening effect of cold-working is due to strain in atomic structure. Another way to accomplish such atomic strain is by *alloying*. When it is considered that each metal has a characteristic space-lattice and a different atomic size, it can easily be understood how the presence of stranger atoms in the atomic arrangement of any metal will set up a strain. That is the theory of hardening and strengthening by alloying.

Before studying some of the most common nonferrous alloys, the student should review Chap. 4 in regard to constitution diagrams.

Peritectic Diagrams

For a better understanding of the alloys in this chapter at least one more case is needed. This is the most difficult one.

It is usually called the *peritectic* reaction. The "Metals Handbook"[1] defines the peritectic reaction as "an isothermal reversible reaction in binary alloy systems in which a solid and a liquid phase react during cooling to form a second solid phase." Here it will be called Case V and explained in a general way with the help of Fig. 102.

Suppose we have two metals, *M* and *N*, that react to form the diagram in Fig. 102. Reactions, on cooling, of compo-

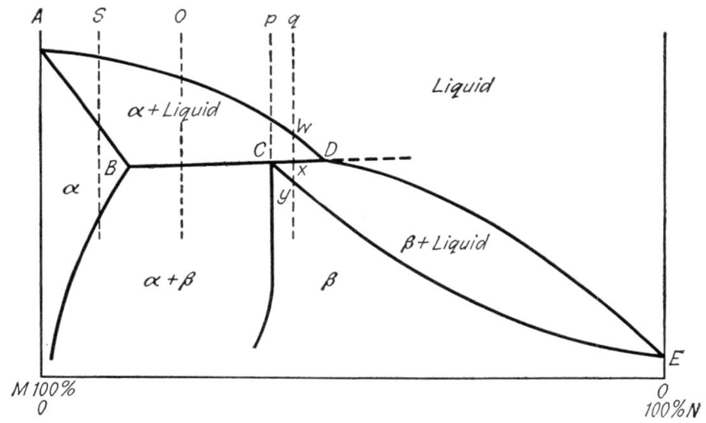

FIG. 102. Diagram illustrating Case V.

sitions *s*, *o*, and *p* are already familiar to us. It is when we come to an alloy of composition *q* that we encounter a new situation. We can calculate the liquid and solid constituents in cooling to *BCD* (by our principle of horizontals). In fact, the principle becomes very useful when we reach that line. If there were a eutectic point along the line instead of a peritectic, everything below the line would be solid.

But in this case, instead of extending as a straight line (dotted) it drops down, and we have a liquid phase to deal with

[1] "Metals Handbook," p. 10, American Society for Metals, Cleveland, 1948.

when we construct our horizontals (intersecting DE). This is because either some solid solution or some intermetallic compound is not stable enough with respect to the liquid to remain solid. (The above figure represents the case of a solid solution.) When the alloy of composition q passes the line BCD in cooling, it is confronted with a composition decided by horizontals drawn between CE and DE (and, of course, it finally reaches y and becomes β solid solution).

One question may arise here. What has become of the α solid solution, which solidified between w and x? The answer is that while the alloy remained at constant temperature at the BCD line (a characteristic that eutectics and peritectics have in common), a change was taking place. There was not a great deal of α solid, but what there was absorbed enough metal N to become β solution. This reasoning presents no difficulty when we consider that the liquid part between w and x already contained this excess of N metal. It might be noted that an alloy of composition p would have changed entirely to β solid solution during its pause at the BCD temperature line. Compositions s and o can be readily traced.

NOTE: More detailed descriptions of the peritectic reaction and constitution diagrams in general can be found in good texts on metallurgy, especially Gilbert E. Doan and Elbert M. Mahla, "Principles of Physical Metallurgy," McGraw-Hill, and the "Metals Handbook" of the American Society for Metals.

COPPER-BASE ALLOYS

Copper

Copper is a red metal with a melting point of 1981°F. It comes to our attention most frequently in the electrical industry, for it is exceeded only by silver in its electrical conductivity. But copper and its alloys are also remarkable for toughness and strength, so much so that at a much higher price it competes with iron and steel, where ductility, malleability, and resistance to corrosion are considered. The pure metal can vary from 30,000 to 70,000 p.s.i. in tensile strength according to the degree of cold-working. The per cent elongation for soft copper can be 45.

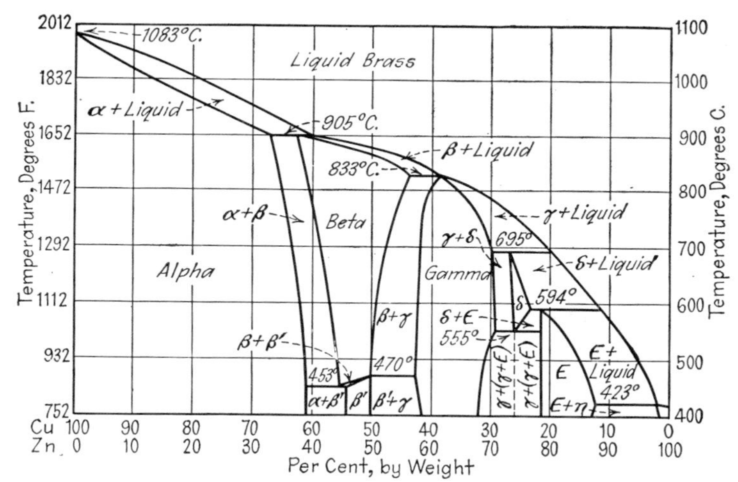

FIG. 103. Copper-zinc constitutional diagram. (*Drawn by A. Phillips.*)

Zinc

Zinc is a white metal with a bluish cast, melting at 787°F. As ordinarily seen in the cast condition, it is a very brittle and coarsely crystalline metal, but it is quite malleable between 200 and 300°F. Its principal uses are in alloys as a protective coating (galvanizing) and in zinc die castings (see page 208).

Brasses

The discussion in the last section paved the way for an appreciation of the copper-zinc diagram (see Fig. 103). The only alloys of commercial importance, however, are those with at least 50 per cent copper. (Above 50 per cent zinc makes them so brittle that they find no use except for decorative purposes.) This limits the commercially important brasses to α brass from 0 to 39 per cent zinc; $\alpha + \beta$, 39 per cent to about 45½ per cent zinc; and β brass up to about 50 per cent zinc. These can be checked by the diagram.

Brasses are used instead of steel because of their corrosion resistance, formability, adaptability to machining, and better casting quality. The latter depends somewhat on the lower temperature required for melting brass.

With their large range of composition, brasses vary considerably in formability by hot- and cold-working. The following table indicates how annealing and cold-working can affect the properties of a 75 per cent copper–25 per cent zinc alloy:

Type	Tensile strength, p.s.i.	Elongation, per cent
Light anneal	52,000	50
Soft anneal	45,000	62
Sheet spring	92,000	3
Wire spring	125,000	

An important application of brass is its use for cartridge cases. The composition is 73 to 66 per cent copper and 27 to 34 per cent zinc. Its advantage here is its formability.

Not until the Second World War was the art of forming steel sufficiently advanced so that steel could replace the more expensive brass in cartridge cases.

Brasses have a pleasing color. With 10 per cent zinc added, a true bronze color is obtained; 20 per cent zinc produces a red-gold color; and above 25 per cent the typical yellow-brass color begins.

Manganese "Bronze"

The addition of a third element to a binary alloy often imparts greater strength. An example is the addition of a little manga-

with lead contents up to 25 per cent will be discussed under *bearing metals.*

Aluminum Bronze

This is a very popular casting metal. Some of its desirable qualities that determine its use are the following: corrosion resistance, toughness, low coefficient of friction, and fair electrical conductivity. Thus it is adapted to a variety of uses, such as motor and generator brush holders, marine pumps, propellers, pickling crates, gas-stove grill plates, gear wheels and worm wheels, just to mention a few. Usually this metal is cast in permanent molds, to take advantage of "chilling" (see Figs. 84, 85, and 86). Several compositions are "heat-treatable."

Copper-Silicon

These alloys are noted for their corrosion resistance. Trade names are Duronze, Herculoy, and Everdur. The last has about 4 per cent silicon and 1 per cent manganese. All three have the strength of steel and can be cast and worked hot or cold.

Beryllium-Copper

This is one of the newer copper alloys and a very remarkable one. Because of its excellent formability in the soft condition, and high proportional limit, great creep resistance, and high fatigue strength in the hardened condition, coupled with good corrosion resistance and electrical conductivity, it finds a multitude of uses such as surgical instruments, rifle parts, diaphragms, nonsparking tools, and especially as springs. It has become known as "the metal which never tires."

The addition of 1 to 2.25 per cent beryllium to copper produces a metal that has a tensil strength of 60,000 p.s.i. in the annealed condition, which by cold-rolling can be increased to

100,000 p.s.i. Remember that pure copper has a tensile strength of about 30,000. By proper heat-treatment the strength of beryllium-copper can be increased to even 200,000 p.s.i. This heat-treatment is known as *age-hardening, aging,* or *precipitation-hardening.*

AGE-HARDENING ALLOYS

Certain types of alloys such as copper-beryllium, duralumin, and others (see page 197) when heated to an elevated temperature, quenched, and then held for a time at some certain lower temperature, show definite changes in properties with time.[1] Hence the terms *aging* and *age-hardening.* The term *precipitation-hardening* will be understood as this discussion proceeds.

Aging will occur only with those alloys of the solid-solution type in which the solubility of the dissolved constituent decreases with falling temperature. This condition is shown in Fig. 105, in which *DBC* is the solid-solubility curve and Z_1 and Z_2 represent typical age-hardenable alloys. At room temperature these alloys consist of two phases in equilibrium: a saturated solid solution of composition *D* and another constituent of composition θ. The second phase may be a pure metal, an intermetallic compound, or another solid solution. On heating the alloy Z_1 to a temperature *T*, all of the constituent θ goes into solution. If the alloy is then rapidly cooled to some low temperature, say room temperature, it will consist of a supersaturated solution of composition Z_1, which became unstable with respect to θ when it crossed the line *DC*.

In the case of alloy Z_2 heated to temperature *T*, the phase θ does not completely dissolve, since the maximum solid solubility

[1] WILLIAMS and HOMERSBERG, *op. cit.,* pp. 119–120. Used by permission.

at T is B. However, here again, on quenching to room temperature, the solid solution is supersaturated with respect to θ. (The student can check these statements by the principle of horizontals referring to the line DC for equilibrium compositions.) The degree of supersaturation is obviously greater in alloy Z_2 than in Z_1; furthermore, the undissolved particles of θ may be helpful in inhibiting grain growth. On the other hand,

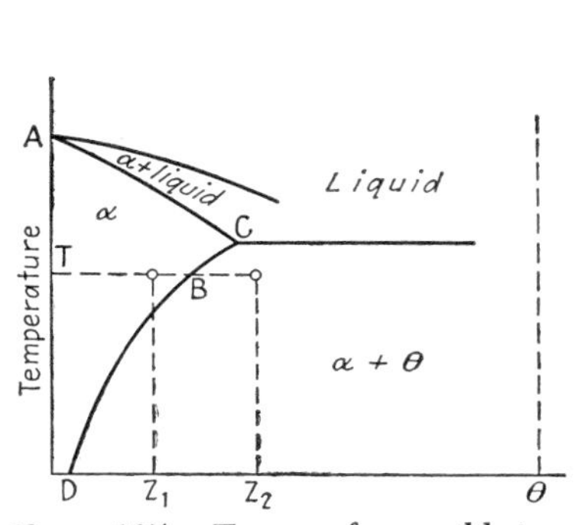

FIG. 105. Type of equilibrium diagram of alloys susceptible to age-hardening. (*Reprinted by permission from Williams and Homerberg, "Principles of Metallography," 5th ed., McGraw-Hill,* 1948.)

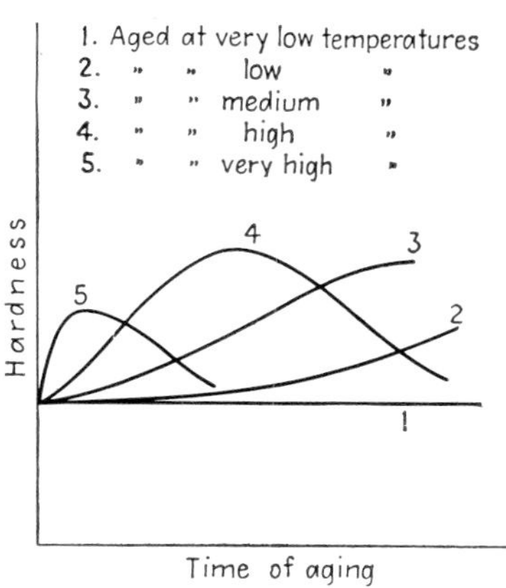

FIG. 106. Age-hardening curves for a hypothetical alloy. (*Reprinted by permission from Williams and Homerberg, "Principles of Metallography," 5th ed., McGraw-Hill,* 1948.)

Z_1 is less likely to undergo incipient melting during forging or heat-treatment (Z_2 is nearer the liquidus line).

The solution heat-treatment described above is an essential part of the age-hardening operation and results in an unstable solid solution that tends to decompose during aging by precipitating θ until the composition of the solution attains the equilibrium value D. The relation of temperature toward the attainment of this equilibrium is illustrated by Fig. 106. At very low temperatures no decomposition occurs. As the aging temperature is raised, however, there is an increase in hardness up to a certain limit, above which hardness decreases again. The latter stage is known as *overaging*. Apparently, at least in the case of aluminum and copper, the greatest strain in the

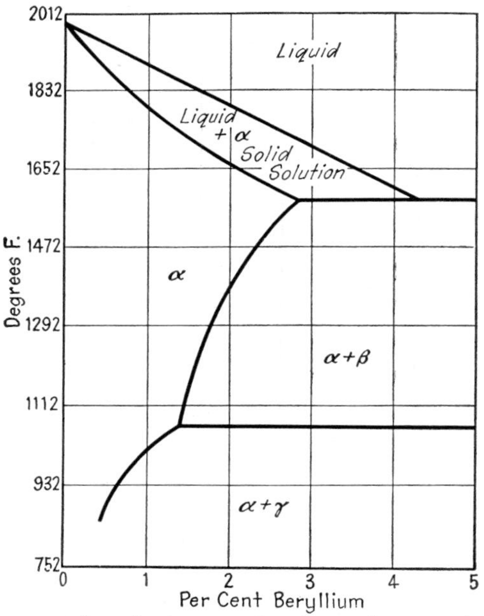

Fig. 107. Copper-rich end of copper-beryllium diagram showing precipitation-hardening possibilities. (*From* Doan, "*Principles of Physical Metallurgy.*")

atomic structure comes when the dispersion of θ ($CuAl_2$ in the case of copper-aluminum) is still of submicroscopic size. According to Guinier and Preston, the hardness is due to a preparation for precipitation within the matrix lattice rather than to the formation of discreet particles. In other words, as soon as particles are formed and ejected, some of the tension is relieved. Thus, age-hardening is another example of hardening resulting from strained atomic structure.

Of all the precipitation-hardening alloys, the aluminum- and copper-base alloys are the most common. However, many more applications of the principle may be found. An alloy of 92½ per cent silver and 7½ per cent copper is a good example.

A part of the copper-beryllium diagram is here given to show its application. After being worked, the metal is quenched at 1340°F. and then reheated to somewhere between 450°F. and 600°F.

One application of hardened beryllium copper is to tools, such as cold chisels, which can be used in inflammable atmospheres, where the carbon in carbon-steel tools would throw sparks. Beryllium copper has another advantage. It is a fine casting metal because it shrinks so little. The production of small parts, such as rifle parts, can be speeded up because no finish-machining is necessary.

ALUMINUM AND ALUMINUM-BASE ALLOYS

Aluminum

Since man has taken wings, aluminum has become one of the most important metals. Unalloyed it is used for electrical conductors, in the thermite welding process, and for thermite incendiary bombs. Aluminum is a weak metal if unalloyed, developing only 14,000 p.s.i. tensile strength.[1] Alloy No. 12, with about 8 per cent copper, is used for crankcases, oil pans, transmission housings, cooking utensils, etc. This can be welded.

Aluminum is the lightest metal in common use with the exception of magnesium. It also has another valuable quality,

[1] This refers to ordinary purity; 99.99 per cent aluminum stands only 9,000 p.s.i. but has an elongation of over 50 per cent.

its heat conductivity. And it is more because of this quality than because of its lightness that it finds use in the manufacture of pistons for automobile and aircraft engines. It makes the cooling of the motor less difficult. Most pistons are permanent-mold castings containing 10 per cent copper, 1¼ per cent iron, and ¼ per cent magnesium. They are used in the heat-treated condition.

Duralumin

When the first Zeppelin dirigibles were being made in Germany, in 1906, Alfred Wilson[1] discovered the heat-treatment that we now call *precipitation-hardening,* by experimenting with an

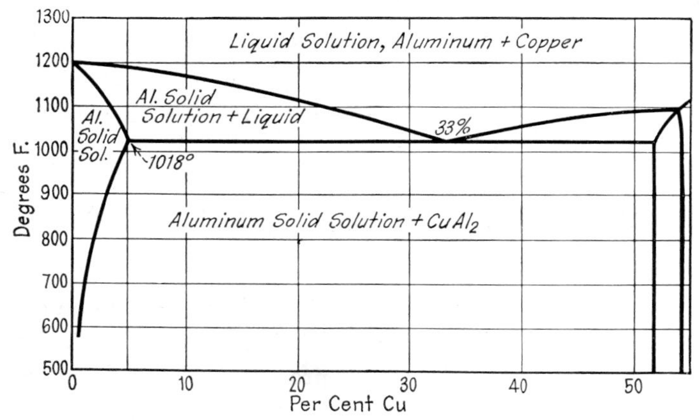

FIG. 108*a*. Part of constitution diagram for copper-aluminum alloys.

alloy of 94 to 95 per cent aluminum, 4 per cent copper, and about ½ per cent each of manganese and magnesium. Thus duralumin has become a classic example of this method of hardening, and its use has wonderfully helped in developing the airplane and even the streamlined train.

[1] The real explanation of the phenomenon was given by Merica, Waltenberg, and Scott, *U.S. Bur. Standards Sci. Paper* 347, 1919.

The heat-treatment of duralumin may be seen from Fig. 108*a*. Suppose an alloy, 96 per cent aluminum and 4 per cent copper, is heated to about 1020°F. and held until in equilibrium. The diagram shows that it is a solid solution. Upon quenching from here, the fast cooling prevents the separation of copper (as $CuAl_2$), which would be natural upon slow cooling. Instead the alloy exists as a supersaturated solution; if given time, the $CuAl_2$ (about 53 per cent copper and 47 per cent aluminum)

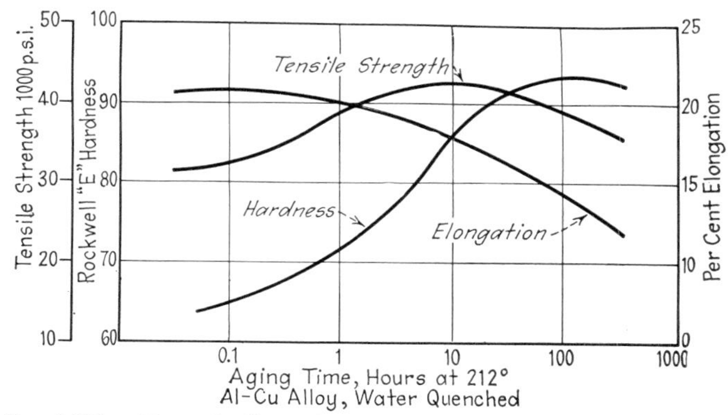

FIG. 108*b*. Typical effect of time on precipitation heat-treatment.

will precipiate.[1] Heating increases the rate of precipitation, as shown in Fig. 108*b*. Note that 10 hr. is the ideal time for aging at 212°F., because longer time will result in a decrease in tensile strength.

After development in the airplane industry, these aluminum alloys have been used for a great number of structural purposes where lightness is required, as in bridges.

An increasingly important group of casting alloys contains from 5 to 15 per cent silicon. Their fluidity is high, and so

[1] Some aluminum alloys age-harden so easily that if it is desired to form them, they must either be formed immediately or stored at subzero temperature.

they can be cast into intricate molds, and, like all silicon alloys, they offer high resistance to corrosion. Their strength is not quite so high as aluminum-copper alloys.

The discussion of aluminum leads naturally to still lighter alloys, now coming into prominence.

MAGNESIUM

Magnesium has a specific gravity of 1.74, as compared with aluminum, 2.70.[1] It is never used alone, being usually alloyed with aluminum (and often with small amounts of manganese). As the aluminum-magnesium diagram has areas of decreasing solubility in the solid state (see Case III), many of the alloys at each end of the diagram are capable of precipitation-hardening.

Dowmetal

One of the most famous is known as Dowmetal,[2] the composition of which, familiarly known as SAE 50, is about 6 per cent aluminum, 3 per cent zinc, less than ½ per cent manganese, and the balance magnesium. The heat-treatment consists of holding the castings for solution at 715 to 730°F., then air cooling by forced draft in the furnace, followed by aging (precipitation) treatment at 350°F. (time according to properties desired).[3] Size for size, test bars and castings of Dowmetal do not show to their full advantage. It is when they are compared weight for weight with other materials that the value of this metal can be realized. The following tables[4] show this in a startling manner.

[1] Also compare 7.1 for zinc, 7.9 for steel, and 8.5 for brass.

[2] Named by its developers, The Dow Chemical Company, who extract metallic magnesium from such salts as the sulfate (epsom salts) obtained from wells and the ocean.

[3] EDWIN F. CONE, The Ford Magnesium Foundry, *Metals and Alloys*, March, 1942, p. 397. Very instructive and should be read.

[4] NORMAN E. WOLDMAN, Magnesium in Aircraft, *Metals and Alloys*, October, 1940, p. 430. Used by permission of the publishers.

TABLE 1. COMPARATIVE PROPERTIES OF DOWMETAL VS. OTHER METALS —EQUAL-VOLUME RELATIONS

Name	Relative weight	Tensile, strength, p.s.i.	Elongation, per cent	Fatigue endurance limit in bending
Mild steel	4.4	60,000	30	30,000
Alloy steel	4.4	100,000	20	50,000
Aluminum alloy, cast, heat-treated	1.6	33,000	8	
Duralumin	1.6	60,000	20	15,000
Dowmetal, cast, heat-treated	1.0	33,000	10	9,000
Dowmetal, wrought, heat-treated	1.0	42,000	12	14,000

TABLE 2. COMPARATIVE PROPERTIES OF DOWMETAL VS. OTHER METALS —EQUAL-WEIGHT RELATIONS

Name	Relative weight	Tensile strength, p.s.i.	Fatigue endurance limit in bending
Mild steel	4.4	60,000	30,000
Alloy steel	4.4	100,000	50,000
Aluminum alloy, cast, heat-treated	4.4	91,000	
Duralumin	4.4	172,000	70,000
Dowmetal, cast, heat-treated	4.4	145,000	80,000
Dowmetal, wrought, heat-treated	4.4	185,000	125,000

Considering the combustibility of magnesium, as evidenced by magnesium ribbon in flashlight bulbs and magnesium cases for thermite bombs, it is evident that special precautions are necessary in melting and casting magnesium alloys. One requirement in regard to sand-mold casting is the addition of some chemical to the sand mix that will prevent the action of moisture on the hot metal.

Aging in Steel

Now that we understand the principles of precipitation-hardening, we can take up an interesting characteristic of very low-carbon steels. Referring to the iron-iron carbide diagram (see page 69), it will be noticed that carbon is slightly soluble in alpha iron, about 0.035 per cent at the A_1 line, and only 0.008 per cent at ordinary temperatures. Thus it may be seen that in a very low-carbon steel, say 0.05 per cent, there is a tendency to throw carbon out of solution when cooling from the critical temperature. Even if the carbon is not actually thrown out of combination, there is at least atomic stress which causes hardening. Sheets of very low-carbon rimmed steel, if held in storage for months, are liable to become so brittle as to greatly hinder forming or punching unless annealing is resorted to. The remarkable fact in connection with such aging is that it does not occur in steels of 0.10 per cent carbon or higher. A possible explanation is that those steels possess enough pearlite to attract any carbon that is being thrown out.

BEARING METALS

Bearing metals can be divided into five classes:

1. The leaded bronzes, already mentioned, containing from 15 to 25 per cent lead, and in extreme cases, 30 per cent. Those with the highest copper content, say 80 per cent copper, 10 per cent tin, and 10 per cent lead, are used for carrying heavy loads, while those of, say 70 per cent copper, 5 per cent tin, and 25 per cent lead, are used for lighter loads and faster speeds.

2. The tin-base bearing alloys. These are the *babbitts* and were the most widely used up to the time that tin became scarce. They range from 90 per cent tin, balance copper and antimony, to 65 per cent tin, 15 per cent antimony, 2 per cent copper, and 18 per cent lead.

3. The lead-base babbitts. They range from 85 to 75 per cent lead, balance antimony and tin. These bearings carry only light loads at moderate speeds, but a modification with 1 to 3 per cent arsenic in addition to the above will carry heavy loads. A less-used lead-base metal makes use of calcium as a hardener for the lead.

4. The cadmium-base metal. These make good bearings and would probably be used more if cadmium were more plentiful.

5. Copper-lead bearings are used largely in the automotive field.

A modern trend is toward a comparatively thick backing of some hard metal such as bronze with a thin inner lining of a soft metal such as babbitt, silver, or indium. It can be seen that a silver-lined, steel-backed bearing would withstand a much higher temperature before failing than would a babbitt bearing—with the melting point of silver at 1760°F., while tin melts at 450°F.

LEAD AND LEAD ALLOYS

The importance of lead in bearing metals was brought out under the previous heading. Lead does not mix well with copper, so such alloys are often melted in some form of furnace where the charge can be kept in motion, for example, the Detroit Rocking Electric Furnace (see Fig. 109).

Lead is a good example of the use of a metal because of special qualities. With all the striving for saving in weight in aircraft through using aluminum and magnesium in construction and even using tires inflated with helium, the storage batteries, still have the same heavy lead plates common to all storage batteries, and the gasoline is weighted down with tetraethyl lead for its antiknock qualities.

Mention was made of calcium-hardened lead as a bearing

FIG. 109. Tapping a heat of bronze alloy from a 175-kw, 500-lb. Detroit Rocking Electric Furnace. Heat is supplied by two large carbon electrodes —one entering at each side. A 700-lb. heat of bronze can be tapped every 40 min. (*Courtesy Kuhlman Electric Furnace Company.*)

metal. This is a precipitation-hardening alloy and is being tried out as a substitute for antimony for such uses as cable sheath and type metal. Standard type metal contains about 4 per cent tin, 11 per cent antimony, and the balance lead.

Solder

After all the attempts at compounding a low-melting solder without the use of tin, lead-tin solder[1] is still the favorite.

[1] The term *solder* commonly refers to alloys with a melting point of less than 700°F. These alloys are used to join other metals. Often they are called *soft* solders to distinguish them from *hard* solders, or *brazing* alloys, which melt around 1300°F. Welding, of course, refers to joining metals at their melting point or above.

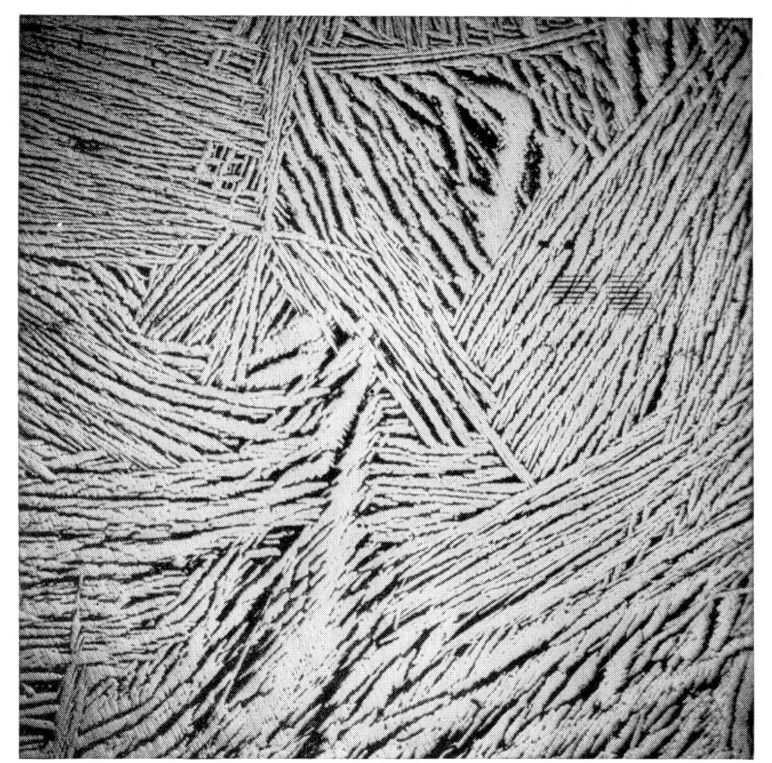

FIG. 110. Photomicrograph of lead-tin eutectic showing the resemblance to the pearlite eutectoid of Fig. 34. Practically all eutectics show this structure. (*Courtesy General Electric Company.*)

Common lead-tin solder is about half and half. The solder that the plumber uses for "wiping" joints contains a higher percentage of lead. As may be seen in the diagram, this allows a greater range of temperature during which the solder is in the mushy stage, thus allowing time for the plumber to form the joint before solidification sets in. It is interesting to note that a large pot of tin-lead solder should be stirred while cooling, or else upon solidification the heavier lead crystals will sink to the bottom and the eutectic crystals tend to rise.

In order to conserve tin, a series of soft solders of low tin content are being worked out, the composition of which may be 70 to 90 per cent lead, 10 to 25 per cent tin, 1½ to 2¼ per cent silver, and, in some, 5 per cent bismuth. Brazing alloys, *hard solders,* melting at red heat, get their name from the brass

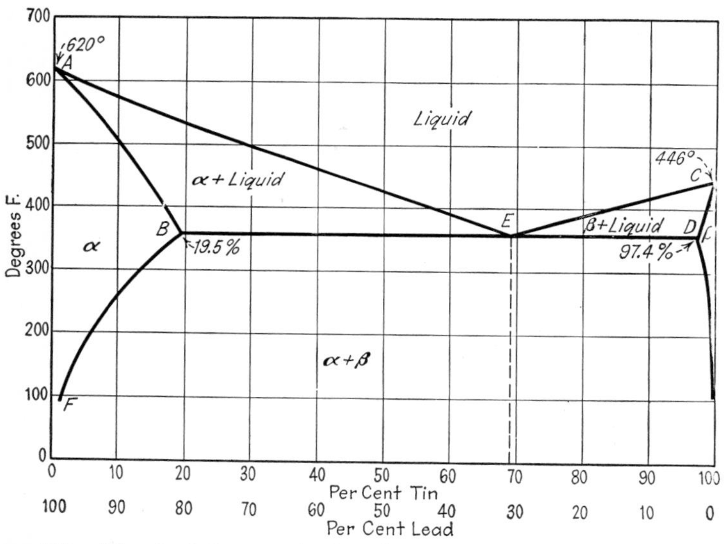

FIG. 111. The lead-tin constitution diagram, an example of Case III. Note that pure lead melts at 620°F. and pure tin at 446°F. but that additions of one to the other lower its melting point. Solidification is complete below *ABEDC*. *FB* marks a transformation within the solid, corresponding to *DC* in Fig. 105.

(58 per cent copper, 42 per cent zinc) that is most commonly used for brazing. However, a *silver solder,* 35 per cent copper, 49 per cent silver, and 16 per cent zinc, is often used because it has a somewhat lower melting point. It is unfortunate that no solders have been devised with melting points intermediate between the *soft* and *hard* series.

Low-melting Alloys

It is evident that alloys of low melting points are often desired. Let us take an extreme case. Consider the eutectic point of lead-tin diagram (see Fig. 111). The melting temperature of this alloy is 356°F.—much lower than either of the metals alone. Now let us lower this still more by adding a third metal and getting the triple eutectic point. Then, adding a fourth, we have the metal used in the plugs of sprinkler systems. Several combinations are used. Lipowitz's alloy, 50 per cent bismuth, 10 per cent cadmium, 27 per cent lead, and 13 per cent tin, melts at 150°F.

NICKEL ALLOYS

Nickel has been mentioned as an alloying element in iron. It has many uses in the pure state, such as in the Edison storage cell and as a catalyst in the hydrogenation of oils, but the greatest tonnage of nickel is used in alloys. Annealed nickel has a tensile strength of 65,000 to 75,000 p.s.i., with an elongation of 43 to 53 per cent.

Nickel-chromium alloys are widely used in the electrical industry for heating elements in toasters, flatirons, percolators, etc. Such apparatus use an alloy of 62 per cent nickel, 15 per cent chromium, and the balance iron, and have a melting point of 2462°F. For higher temperatures, the 80-20-nickel–chromium alloy is used. It has a melting point of 2550°F.

Among other corrosion-resisting nickel alloys might be mentioned Hastelloy C, containing nickel; molybdenum; chromium; tungsten and iron; and Ilium G, which contains nickel chromium, copper, molybdenum, and iron.

The nickel-copper alloys are of great importance for use as corrosion-resistant metals of high strength.

Monel Metal

Monel metal is expensive, but it is practically noncorrodible and is used largely for propeller blades, steam pressure valves, and steam-turbine blading. It consists of 68 per cent Ni and 27 per cent Cu, with the 5 per cent balance Fe, Mn, and Si. It has a tensile strength of 70,000 to 85,000 p.s.i. and 35 to 50 per cent elongation, but it can be worked to as high as 140,000 p.s.i. and 2 to 10 per cent elongation. As neither element forms carbides, and as the constitution diagram resembles that of gold and silver, its hardness can be increased only by cold-working The addition of nickel to copper rapidly removes the red color of copper, the 15 per cent alloy having a faint pink cast, while alloys with 20 per cent or more nickel are essentially white. The United States 5-cent coin is 25 per cent nickel and 75 per cent copper. *Nickel silver* contains nickel and copper in about the same proportion that they occur in Monel metal, plus about 17 per cent zinc. This alloy resembles silver and is used for such purposes as zippers and as base for silver plating.

ZINC AND ZINC ALLOYS

The use of zinc as a constituent of brass has been mentioned. Its most extensive use is for galvanizing, but its use in alloys for die castings is growing in importance. Zinc is a very coarsely crystalline and brittle metal when cast, but wrought zinc can have a tensile strength of about 20,000 p.s.i. and 40 per cent elongation.

Die castings

A die casting can be defined as any casting formed under pressure greater than gravity. The zinc alloys used for such pressure casting[1] contain about 3½ per cent aluminum, and

[1] The alloys are all protected by patents.

copper up to 2½ per cent. This metal is well adapted for this purpose, as it melts at about 717°F. and has 40,000 to 48,000-p.s.i. tensile strength and 5 per cent elongation. It is used for radio chassis, carburetors, interior hardware, and many intricate ornamental shapes.

The nonferrous alloys mentioned in this chapter do not begin to cover the field. Just how numerous they are may be understood by the realization that the dentist has a choice among 150 different gold alloys for filling teeth and for bridge work.

QUESTIONS

1. Would you think that a high-lead bronze bearing could be centrifugally cast?
2. Since studying about decrease in solid solubility (see Fig. 105), can you find a similar area in the iron-iron carbide diagram (Fig. 31)? Explain it.
3. Why is a carbon-steel cold chisel liable to emit sparks when one made of beryllium copper will not?
4. Name some substitutes for tin in alloys.
5. What is precipitation-hardening?
6. Name and describe four other hardening methods.
7. List six examples of metals capable of precipitation-hardening or -aging, giving the base metal and its alloy or alloys.
8. Do any of the metals mentioned in this chapter show any hardening ability with the addition of carbon?
9. Name three beneficial effects of lead as alloy.
10. Name three other uses of lead aside from alloy.

REFERENCES

Brick, R. M., and Arthur Phillips, "Structure and Properties of Alloys," 2d ed., McGraw-Hill, New York, 1942.

Ellis, Owen, W., "Copper and Copper Alloys," American Society for Metals, Cleveland, 1947.

Gillett, Russell, and Dayton, "Bearing Metals," *Metals and Alloys*, September, October, November, and December, 1940.

"Metals Handbook," American Society for Metals, Cleveland, 1948.

Von Zeerleder, Alfred, "Technology of the Light Metals," Elsevier Publishing Company, New York, 1949.

Chapter 12

TESTING

All materials purchased by any large manufacturing concern are tested before they are accepted, to see whether they are suitable for the purpose intended. This is especially true of metals. Sometimes, if the metal is intended for special uses, these tests are very elaborate. Usually a careful chemical analysis is made, because, while such an analysis does not give the actual physical properties, it tells what can be expected of the metal, and it usually shows the reason for any metal failing to pass physical tests. The importance of the miscroscope has been emphasized all through this text.

Mechanical Tests

The most important tests used in the purchase of metals however, are the mechanical ones. The hardness tests, such as the Brinell and the Rockwell, the tensile and impact tests, and the measurement of elongation, etc., are so definite that they can be made the basis of specifications and really give information as to whether the metal is adapted to the desired use. The Brinell and Rockwell tests for hardness are perhaps the most used of any tests because they can be made without destroying the part tested.

Brinell Hardness Tester

The Brinell instrument presses a 10-mm. (0.394-in.) chrome-steel ball into the material with a load of 3,000 kg. (6,600 lb.). A so-called *Brinell hardness number* (Bhn) is calculated from the diameter of the impression made by the ball according to the formula:

$$\text{Bhn} = \frac{P}{\frac{\pi D}{2}\left(D - \sqrt{D^2 - d^2}\right)}$$

where P = the load, 3,000 kg.
D = diameter of the ball, 10 mm.
d = diameter of the impression in millimeters

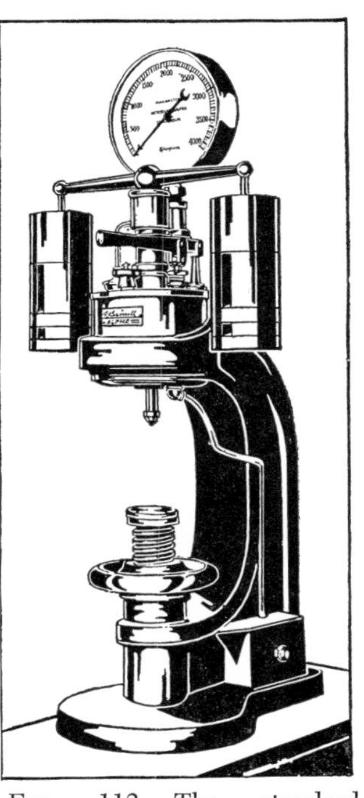

FIG. 112. The standard Brinell hardness tester. (*Courtesy Herman A. Holz.*)

Rockwell Hardness Tester

The Rockwell hardness tester measures the difference in depth caused by two different loads, a major and a minor load. The penetrator is a diamond cone with rounded point for hard steels, a 1/16-in. steel ball for soft steels, and other sizes for various other materials. The results of each penetrator and load combination are reported on separate scales, designated by letters. In using the machine, a minor load of 10 kg.[1] is first applied and the dial set at zero. In the case of hard steels, a load of 150 kg. is then applied, and the depth of impression as read on the black figures

[1] A kilogram is equal to 2.2 lb.

is reported as *Rockwell C* hardness. In the case of soft steels, the applied load is 100 kg., and the depth of impression as read on the red figures is reported as *Rockwell B* hardness. A variety of penetrators and scales makes this a comprehensive instrument.

It is important to know that the direct reading of impression depth is not the figure used. That would be inaccurate because of the elasticity of both the tested material and of the instrument itself. Such inaccuracy is admirably overcome by first causing a light impression, setting the dial at zero, then putting on a large increment of load, which shows momentarily on the dial. This large increment is then released, but the minor load is retained in order to maintain tight contacts, and the dial now shows the actual depth due to the large increment of load and without either the instrument or the work being under strain. The operations are well explained by the sketches of Fig. 114.

FIG. 113. Rockwell hardness tester. (*Courtesy Wilson Mechanical Instrument Company, Inc.*)

Another Rockwell instrument, known as the *Rockwell superficial-hardness* tester, is adapted to measuring surface hardness of thin sheets, such as razor blades and lightly casehardened surfaces. Similarly, the *Brinell machine,* when used with softer materials, employs a smaller ball and lighter load.

Another hardness tester is the *Scleroscope,* which simply measures the rebound of a diamond-tipped hammer dropped

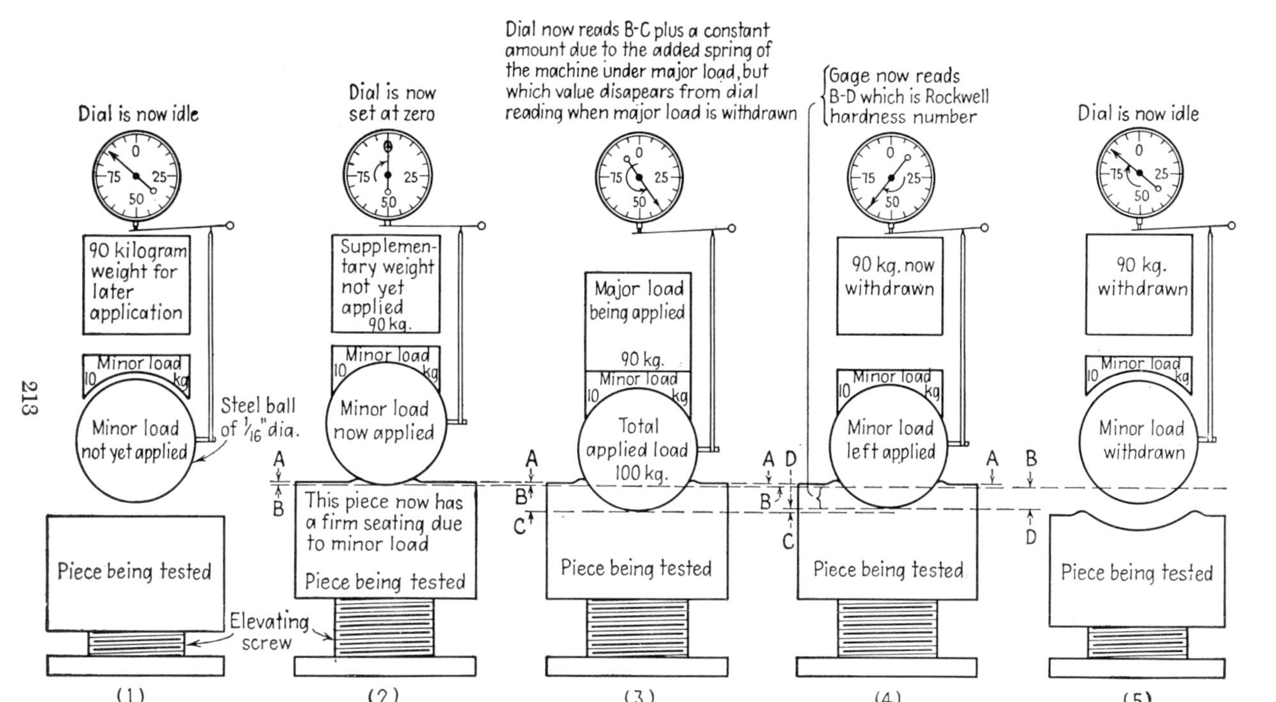

FIG. 114. Illustrating the principle of operation of the Rockwell hardness tester. (*Modified from "Rockwell Hardness Tester," Catalog No. 10, p. 12, Wilson Mechanical Instrument Company, Inc. Reproduced from George F. Kehl, "Principles of Metallographic Laboratory Practice,"* 2d ed., 1949.)

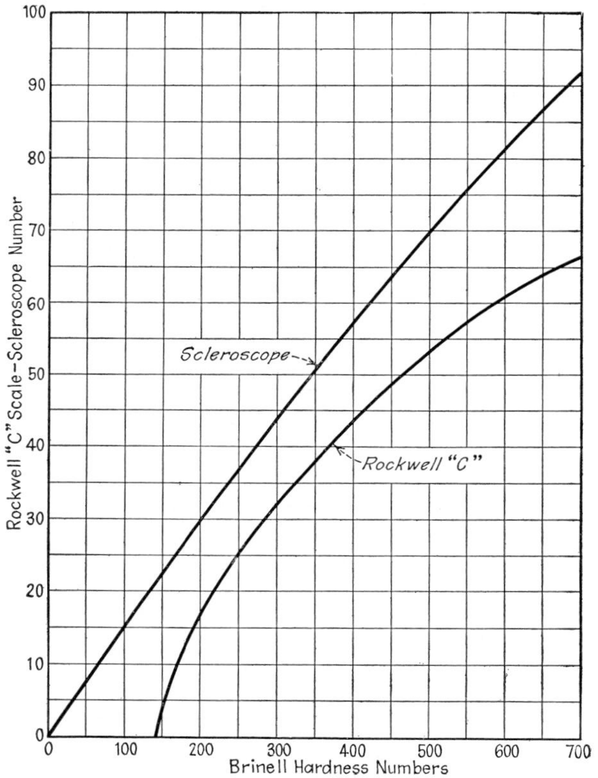

FIG. 115. Conversion chart showing approximate relation between three methods of hardness testing.

from a constant height. Of course, the harder the material, the higher the hammer will "bounce." This instrument is not quite so reliable but has the advantage of portability. A comparison of values by all three instruments is shown in Fig. 115.

The simplest of all hardness tests, of course, is the common file. While its results are never mentioned in specifications, it is extremely handy around the plant. For instance, a fine-toothed file will make no impression on a nitrided surface. A carburized and quenched surface will probably resist the file after a water quench but may show a scratch if oil quenched. Plain-carbon steels are resistant according to their carbon content and heat-treatment.

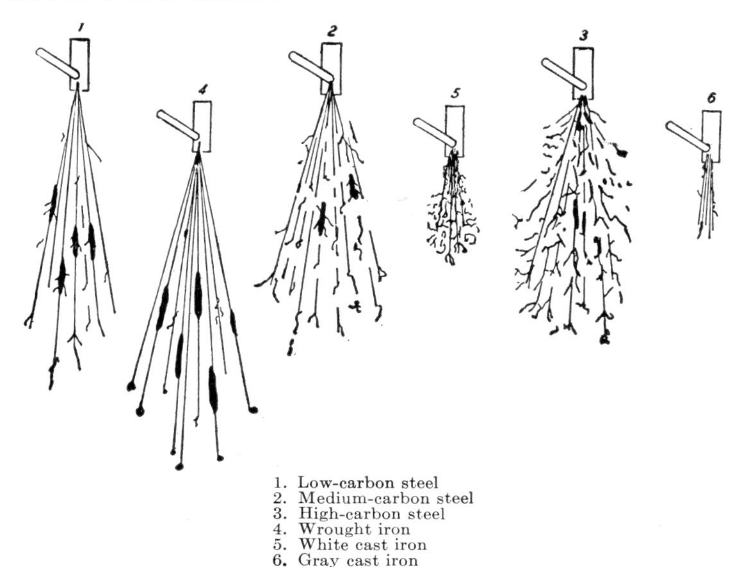

FIG. 116. Characteristic sparks produced by pressing various steels against a swiftly revolving grinding wheel. (*From G. M. Enos, Trans. Am. Soc. Steel Treating, vol.* 12, 1927.)

Spark Tests

Spark tests are very valuable in identifying pieces of steel according to their carbon and alloy content. A little experience will enable anyone to estimate the carbon in plain-carbon steel by noting *stars,* which are white, and forked *rays,* which are a yellow color. Pure iron, held against a grinding wheel, shows only the latter. Carbon can be estimated by the percentage of stars. Tungsten steel causes red sparks; nickel, intense white, etc.

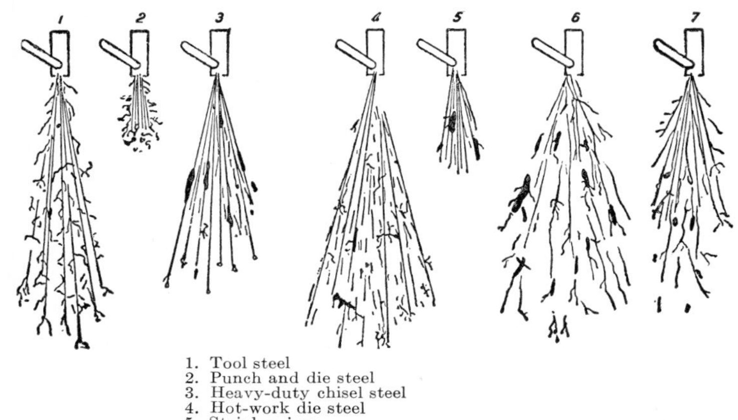

1. Tool steel
2. Punch and die steel
3. Heavy-duty chisel steel
4. Hot-work die steel
5. Stainless iron
6. Nonshrinking tap and threading die steel
7. Magnet steel

FIG. 117. Characteristic sparks produced by pressing various steels against a swiftly revolving grinding wheel. (*From G. M. Enos, Trans. Am. Soc. Steel Treating, vol.* 12, 1927.)

TENSILE TESTS

The most definite of all physical tests are those for tensile strength, per cent elongation, elastic limit, and yield point—all made in one determination by measuring (weighing) the pull[1] (tension) exerted on an accurately dimensioned test piece. Test pieces are machined very accurately (see page 217). They vary in size and shape according to material, but the usual dimensions of the piece used in testing steel are shown in Fig. 118.

The results of the test are always reported in pounds per square inch (p.s.i.), but no bar of exactly 1-sq. in. cross-sectional area is ever used for testing steel because of the enormous load that would be required. The 0.505-in. diameter is chosen be-

[1] In the hydraulic machine the pressure exerted in order to produce the pull is read on a dial.

cause its area is 0.2 sq. in., an easy factor to divide into the results as read.

Tensile Strength

Tensile strength has been frequently mentioned in this book because it is the property most commonly spoken of in connection with the *strength* of a metal. As usually determined in engineering work, it is the maximum stress in a tension test divided by

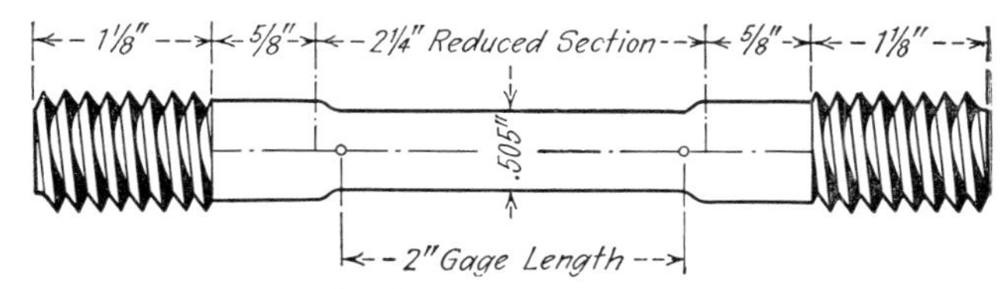

FIG. 118. Standard 0.505-in. tensile test bar for steel.

the original area. However, tensile strength is not the most important quality of steels and many other metals in determining their suitability for use. Long before a test bar breaks, it has *elongated* and lost its original shape.

Elastic Limit

What is very important in most engineering uses is what stress the metal will stand *without deformation* or, more exactly speaking, without *permanent* deformation; for there is a slight deformation with any load. The term *elastic limit* means the maximum unit stress that can be developed in a material without causing a permanent set. The word *elastic* implies a return to original shape when the load is removed, just as would be the case with a rubber band. The true elastic limit is a very difficult point to determine. One would have to apply increasing loads and then release them and measure the piece each time to see if it has returned to its original length. Sometimes this is specified, but it is a tedious operation.

Yield Point

If the specifications are not too exacting, what is known as the *yield point* gives a fairly good idea of what the elastic limit might be. As the name indicates, the test piece has passed the elastic limit and has started to yield (permanent deformation

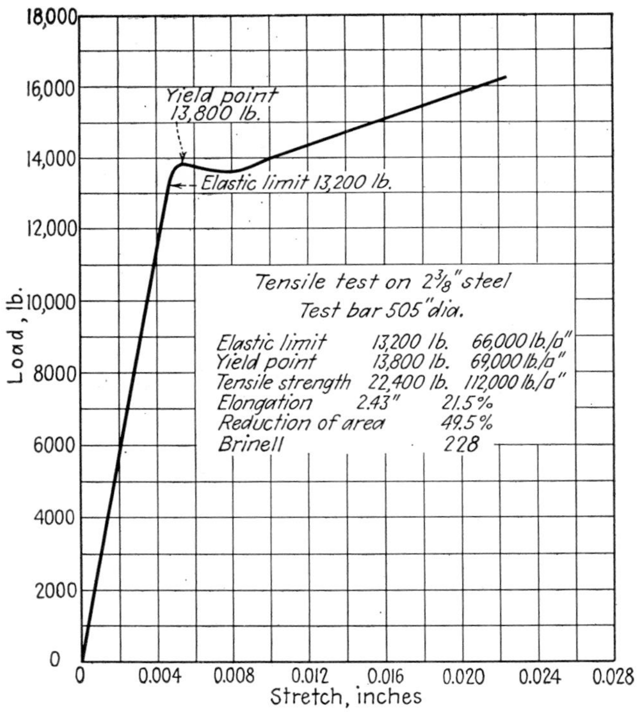

FIG. 119. Copy of an autographic record of a tensile test. (*Courtesy General Electric Company.*)

has set in). One thing that makes the yield point a rather indefinite and unreliable test is the way it is usually determined. The *load,* or *pull,* applied to a bar under test increases so rapidly up to the yield point that in the case of those testing machines where the weight is moved out along the beam by hand, it is almost impossible to stop quickly enough to obtain the exact load where the material begins to yield. Instead the operator is very liable to overshoot the mark, and this will be shown by the drop of the beam;[1] the machine will continue to run, elongating the piece, for several seconds before any increase in load will cause the beam to rise again.

In a way, there is really some justification for the beam dropping, as will be seen on the autographically traced line (Fig. 119). If the operator is careful enough not to overshoot the mark, the test has considerable value in determining the yield point. Machines that trace an autographic record of the test are so expensive that only the largest manufacturers and laboratories can afford them. The Erie General Electric plant has one, and the curve in Fig. 119 is an example of a test made with it. By this method the yield point can be determined accurately. It is the top of the curve just as the piece deforms. The elastic limit is determined accurately enough for practically all specifications; it is taken at the point where the curve first deviates from a straight line.

Scribe Method

In the absence of such an autographic attachment, the elastic limit can be determined by what is known as the *scribe method.* A pair of dividers is set at 2 in., one leg inserted in the lower punch mark (made for measuring elongation later) and a mark inscribed with the upper leg. Suppose specifications called for 60,000 p.s.i. elastic limit; when a load corresponding to that is reached, the load can be released and another mark inscribed with the upper leg of the dividers. If it coincides with the first

[1] *Drop of beam* is a method sometimes allowed in tests where great accuracy is not required. One objection is that it depends too much on the personal equation.

mark, the elastic limit has not been reached. A thickening of the first line is taken to indicate that the limit has just been reached. Two lines would indicate that the limit had been exceeded.

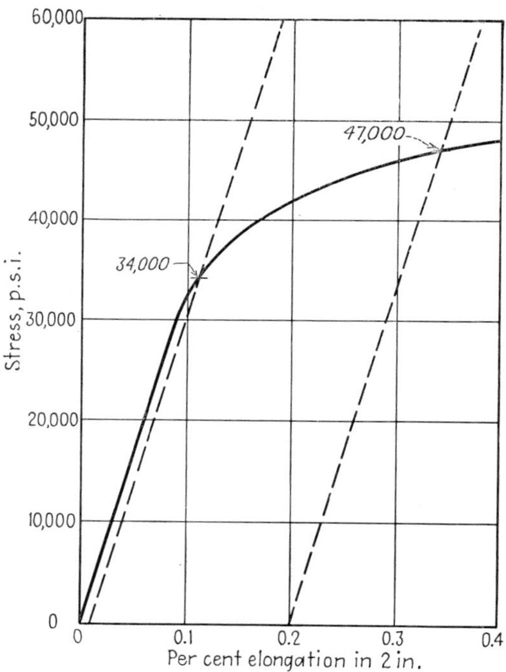

FIG. 120. Illustrating *offset* method of determining "proof stress" and "yield strength."

The methods so far described are applicable to most steels. However, tensile tests on stainless steels and most nonferrous metals result in a graph resembling Fig. 120. It is obviously impossible to determine a yield point on this graph by the methods so far described, and it would be difficult to decide upon a value for the elastic limit.

Nonferrous Tests

In order to make definite and comparable tests, and in order to write values into specifications, for metals showing such tension test curves, what are known as offset values have been agreed upon. For instance, it would be difficult to determine just where the curve of Fig. 120 deviated from a straight line; therefore the straight dashed line is drawn *offset* (set off) 0.01 per cent from the lower part of the curve. The point where this dash line intersects the stress-strain curve (34,000) is the value taken to correspond to an elastic limit; only in this case it has come to be termed *proof stress.* In reporting this value, the amount of offset must also be specified—as "proof stress, 0.01 per cent offset." Similarly, the value corresponding to a yield point in such metals is taken where a line drawn 0.2 per cent offset intersects the stress-strain curve (47,000) and, similarly, is not referred to as "yield point" but as "*yield strength,* 0.2 per cent." The amount of offset must always be included.

Now, in the case of all metals, after the elastic and yield properties have been determined in a tensile test, the load is again applied (usually at a faster rate) until the test piece breaks. The highest load attained is known as the *tensile strength.* It is reported in pounds per square inch based upon the *original area,* which is of course inaccurate (for the area has decreased) but gives a comparative result (see page 223).

Two other properties are then measured, which are indications of ductility. One is the *per cent elongation.* This is measured by setting the broken ends together and measuring the distance between the two punch marks that were originally marked exactly 2 in. apart. The increase in length divided by 2 gives the *per cent elongation in 2 in.* The other property is known as *reduction of area.* As Fig. 121 shows, the broken bar

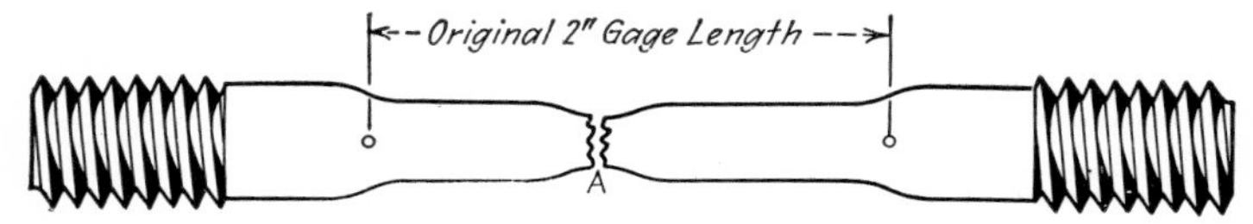

FIG. 121. Tensile test bar after fracture.

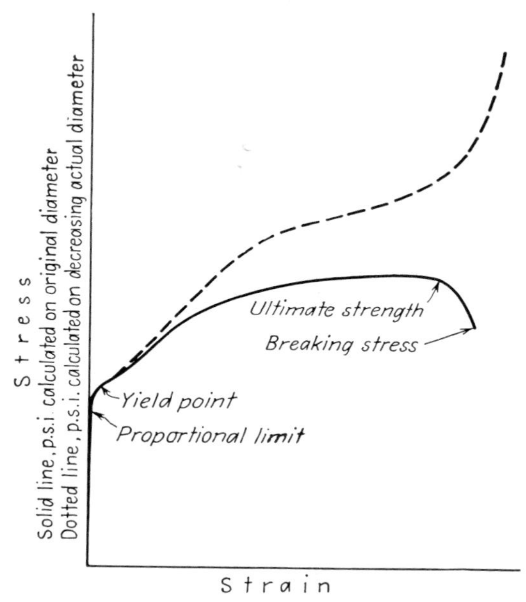

FIG. 122. Sketched completion of test recorded in Fig. 119; also comparison of engineering stress-strain diagram (solid line) with physical stress-strain diagram (dotted line).

has *necked down* at the point of fracture. Percentage reduction of area is the difference between the original area and the final area (point *A*) expressed as a percentage of the original cross-sectional area.

Figure 122 sketches the completion of the test shown in Fig. 119. Obviously the delicate mechanism of the autographic recorder would be injured if left in the test piece up to the breaking point of the piece, so it is always disconnected as soon as the yield point has been passed. Besides, after the yield point has been passed, the further behavior of the metal under stress is not quite so important to the engineer. True, he always wants to know the ultimate strength, but this is usually more for the purpose of comparison than for direct application to any service requirements.

The student of metallurgy, however, is interested in the actual stress, calculated on the progressively decreasing diameter as the test piece is plastically deforming and increasing in length. The curves sketched in Fig. 122 represent a steel that is ductile enough to *neck down* to a small diameter at the point of breaking (*A*, in Fig. 121). Most of this necking down occurred between the point of ultimate strength and the breaking point. (A brittle steel would break at the ultimate strength because it would not neck down.)

The engineering and the physical stress-strain diagrams coincide up to the proportional limit because they are based on the same cross-sectional area up to that point; but beyond that point they begin to separate because they are based on different areas, the engineering on a constant (but false) area corresponding to the original diameter, while the physical is based on the true, decreasing diameter. Where this diameter becomes quite small (as in necking down), there is a wide divergence between the two diagrams.[1]

The student of metallurgy is also interested in the enormous stresses which ductile metals will stand when the pounds per unit area are calculated on the true area at the point of ultimate strength or at the breaking point. As an example, the data given with Fig. 119 show an ultimate strength (commonly known

[1] A good explanation of engineering and physical stress-strain diagrams is given by CLAPP and CLARK, "*Engineering Materials and Processes*," 2d ed., International Textbook, Scranton, Pa., 1949.

as *tensile strength*) of 112,000 p.s.i. and a reduction of area of 50 per cent. Thus the true ultimate strength is 112,000 ÷ ½ = 224,000 p.s.i.

This true ultimate strength really represents the maximum strength that can be developed by cold-working (or by alloying). Up to the elastic limit the small displacement that the atoms suffered would be immediately recovered if the tension were released. However, at the yield point and beyond, permanent slips occurred along planes insides the grains, as roughly sketched in Fig. 78 and well shown in the photomicrograph, Fig. 77.

Of course, finally, grain boundaries and broken grains stopped further slipping, and the piece broke. A test piece stressed to near its breaking point corresponds to a severely cold-worked specimen in that it has lost its ductility—no more elongation or reduction of area is possible.

DYNAMIC AND IMPACT TESTS

The tests considered have been the so-called *static* tests, *i.e.*, tests made comparatively slowly on materials at rest. Other static tests include compression, torsion (twisting), and bending. Creep, already mentioned on page 179, is another. Another class of tests is the *dynamic* tests, made by repeated stresses or sudden shock. Such tests are very valuable in showing how the material will behave in actual use in machine parts. Two such tests will be described.

Fatigue Tests

One type is illustrated by Fig. 123. It can be seen that the load puts the upper surface of the test piece in compression and the lower in tension. The motor revolves the piece as high as 1,500 to 3,000 r.p.m., thus placing every particle of the piece alternately under opposite stresses at every revolution and simulating the stresses which such parts as crankshafts meet in service. Test specimens for fatigue tests must be given a very smooth finish because the slightest scratch will affect the results. This brings up the subject of "notch sensitivity." Often fatigue tests are carried out in duplicate—one bar notched and the other smooth—in order to determine the relative notch sensitivity, which is a very important property to

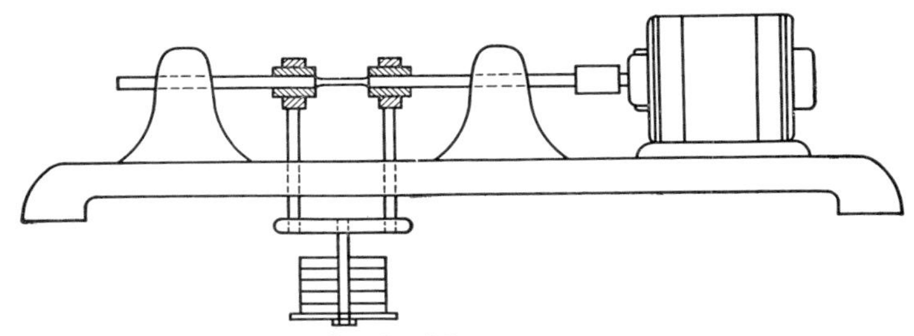

FIG. 123. Sketch of fatigue-testing apparatus.

consider when keyways, for instance, are going to be cut in a shaft. This test is often known as a *fatigue test;* and the point of failure, expressed in load and the number of reverses of stress withstood before failure is called the *endurance limit.*

All fatigue failures occur from *tension* stresses. In fact almost all failures can be attributed to this cause. One reason for this fact is that in almost all cases where compression is involved, the parts can be made large enough and heavy enough to withstand such stresses; that is why the great majority of tests on metals are concerned with tension properties rather than with compression.

Impact Testing

Another type of testing shows the ability of the metal to stand sudden shock or impact. Figure 124*a* shows the dimensions of the test piece, and Fig. 124*b* shows a sketch of a Charpy impact apparatus. With the help of the sketch, it may be seen that the heavy head of the machine swings downward and chops the test piece much as a guillotine would. The tougher and stronger the material is, the more energy it will consume and the shorter distance will the head travel after cutting through.

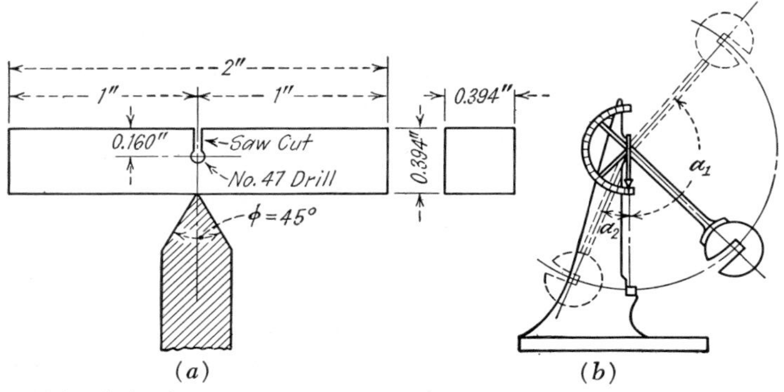

FIG. 124. (*a*) Charpy impact specimen and pendulum cutting face. (*b*) Charpy impact apparatus showing three positions of pendulum. (*From Doan, "Principles of Physical Metallurgy."*)

The reported result is in foot-pounds of energy absorbed by the test piece.[1]

To sum up, the suitability of a metal for its intended purpose is judged by the determination of certain of the following properties:

a. Hardness. Ability to resist plastic deformation.

b. Elastic Limit. Limit of unit strain which it can stand without permanent deformation.

c. Yield Point.[2] In mild- or medium-carbon steels, the stress at which a marked increase in deformation occurs without in-

[1] Measured by the relation of the swings a_1 and a_2, weight of head, etc.

[2] From "Metals Handbook," American Society for Metals, Cleveland, 1948. Used by permission.

crease in load. (Not observed in other steels or in nonferrous metals.)

d. Yield Strength.[1] The stress at which a material exhibits a specified limiting deviation from proportionality of stress to strain.

e. Tensile Strength.[1] The value obtained by dividing the maximum load observed during tensile straining by the cross-sectional area of the specimen before straining. Also called *ultimate strength.*

f. Toughness. Ability to withstand shock or impact.

g. Ductility. Ability to be deformed without rupture. A very valuable property when cold-working and forming is to be considered—measured by per cent elongation and reduction of area.

h. Rigidity, or stiffness. Amount of deflection under load. Its measure is called *Modulus of Elasticity* but is simply the ratio of stress (p.s.i.) to strain (in. per sq. in.). The former are the ordinates and the latter are the abscissas of stress-strain diagrams (see Fig. 119).

i. Fatigue.[1] The tendency of a metal to break under conditions of repeated cyclic stressing considerably below the ultimate tensile strength.

j. Impact. Resistance to sudden shock.

k. and *l.* Torsion and shear strengths are self-explanatory.

m. Creep.[1] The flow or plastic deformation of metals held for long periods of time at stresses lower than the normal yield strength. The effect is particularly important if the temperature of stressing is in the vicinity of the recrystallization temperature of the metal.

n. Soundness. Freedom from flaws (tests for soundness will be considered at the close of this chapter).

[1] From "Metals Handbook," American Society for Metals, Cleveland, 1948. Used by permission.

MICROSCOPIC EXAMINATION

The importance of microscopic examination of metals has been in evidence all through the book. Of the 129 illustrations, 29 are microscopic views. As will be noted, most of these are in the range of 50 to 100 *diameters*, although 1,000-diameter magnifications are often used, and the microscopic range might be said to extend from 10 to 5,000 diameters.

The word *diameters* is used to denote the number of times the diameter of the specimen had been extended, so as not to get the expression confused with the extension of *area*.

Anyone acquainted with photography realizes that the higher the magnification, the more restricted the focusing possible; which means that the surface observed must be polished very evenly, because any slight depression will very likely be out of focus when the main surface is in focus. Besides, "scratch lines" cannot be tolerated anyway.

In preparing a specimen for examination, it is first ground flat—care being taken not to heat the surface, for heat would change the structure. After grinding, the specimen is rubbed to a polish on three or four emery papers of increasing fineness. Each rubbing is done at right angles to the rubbing on the previous paper in order that it can be seen whether the previous scratches have been rubbed out. It might seem superfluous to mention that perfect cleanliness must be observed going from one paper to the next, and yet this care is often not given. Any small particle of the previous size would naturally cause scratches.

After finishing with the finest paper (No. 0000), scratch lines are still visible at most magnifications. These are smoothed away by an operation resembling what a mechanic would term *lapping*. Rapidly revolving disks faced with broad cloth are supplied with a water suspension of rouge or "levigated alumina." This word *levigated* indicates how fine the powder is, for it remains in suspension in water and does not settle. The polish imparted by such fine particles leaves a perfectly smooth surface as far as can be seen by the microscope, except for carbon spots in cast and malleable iron and small inclusions present in most irons and steels. The outline of the grains or their structural nature is not visible.

In order to bring out the grain boundaries and the grain structure, the surface must be treated with one or more of various etching reagents. Some of these are acids. These show the different structures, because acids will attack some structures more readily than others. One etchant, *nital*, a 1 to 5 per cent solution of concentrated nitric acid in alcohol, is used to bring out the boundaries in low-carbon steel, to bring out maximum contrast between pearlite and a cementite network, and among other uses to develop the boundaries of ferrite when present with martensite. Another much used etchant is *picral*, a 4 per cent solution of picric acid in alcohol. It is used to bring out details of structures, and thus helps to identify them. There are a great many other etchants, each serving a specific purpose. Detailed information on all phases of microscopy can be obtained from such books as: R. S. Williams and V. O. Homerberg, "Principles of Metallography," 5th ed., McGraw-Hill, 1948; George L. Kehl, "The Principles of Metallographic Laboratory Practice," 2d ed., McGraw-Hill, 1949.

X-rays

X-rays can be used to show defects in castings, welds, etc., in the same manner that a dentist determines the condition of teeth—only of course with much greater power. Cavities show up as dark spots on the negative because more rays get through at that point. Imperfections deep within the casting darken the negative to the same extent as if they were on the surface.

Steel can be X-rayed successfully through a thickness of 2 in. in less than 10 min., using a million-volt unit. Many airplane and other important parts are put through a 100 per cent X-ray inspection.

Supersonic Vibrations

This is a method of flaw detection which shows great promise. Vibrations in the nature of sound waves but of higher frequency so that they cannot be "heard" by the human ear are used. This method is important because it will penetrate as high as 20 ft. of metal. Within a few inches of the contacting point, a void of 0.02 in. in diameter can be detected; and at a distance of 10 ft., a spherical hole 0.2 in. in diameter will show.

The vibrations are produced electrically and can be anywhere within the range of 50,000 to 5,000,000 per sec. They are electrical impulses that are sent through a quartz crystal. Quartz has the property of expanding and contracting with changes of current. (That was the principle which made the old crystal radio sets work and still has many similar applications.)

The smooth piece of quartz is placed against a smooth surface of the metal with an oil film between to make a good contact. The mechanical waves from the electrically excited quartz are sent into the metal. They travel so much faster in metal than in air that when they reach the other side they are reflected back; or, if they strike a bubble or spot of slag, they are likewise reflected back. The reflected supersonic waves come back to the quartz crystal and are changed into electrical impulses which are caused to register on a cathode-ray oscillograph. This instrument is adjustable, so that the time required for a set of waves reflected from a flaw is shown on the screen, compared with the time required for a set of waves reflected from the far side of the piece under test. Thus the distance to the flaw and its magnitude can be easily read.

Macroscopic Etching

This test is very useful in showing the *fibers* (or the flow of metal resulting from working) in forgings and in rolled products.. It consists of very deep etching by hydrochloric acid, usually warm. *Macro* means large as contrasted with *micro*, meaning very small; and so macroscopic examination refers to what can be seen either by the unaided eye or with the aid of a small glass, up to 10 diameters. Figure 125 is a good example of macroscopic etching, clearly showing the direction of fibers due to forging the head of the bolt.

FIG. 125. An example of macro etching showing fibrous structure due to forging. Bolt was hot forged from 2⅛ in. round. (*Photo by I. C. Mitchell.*)

REFERENCES

AMERICAN SOCIETY FOR TESTING MATERIALS, "Standard Definitions of Terms Relating to Methods of Testing," Standards E6–36, Philadelphia, 1936.

CLAPP, W. H., and D. S. CLARK, "Engineering Materials and Processes," International Textbook, Scranton, Pa., 1949.
KEHL, GEORGE L., "The Principles of Metallographic Laboratory Practice," 3d ed., McGraw-Hill, New York, 1939.
SISCO, FRANK T., "Metallurgy for Engineers," Pitman, New York, 1948.
WILLIAMS, R. S., and V. O. HOMERBERG, "Principles of Metallography," 5th ed., McGraw-Hill, New York, 1948.

APPENDIX

Table 3. Melting Points and Specific Gravities of Some of the Metals

Metal	Symbol	Specific gravity	Melting point, °F.
Aluminum	Al	2.6	1217
Antimony	Sb	6.7	1166
Bismuth	Bi	9.8	520
Cadmium	Cd	8.6	610
Calcium	Ca	1.6	1564
Chromium	Cr	7.0	2822
Cobalt	Co	8.5	2690
Copper	Cu	8.9	1981
Gold	Au	19.26	1945
Iron	Fe	7.9	2800
Lead	Pb	11.4	621
Magnesium	Mg	1.7	1204
Manganese	Mn	7.5	2268
Mercury	Hg	13.5	−37
Molybdenum	Mo	9.0	4748
Nickel	Ni	8.8	2645
Platinum	Pt	21.4	3190
Silver	Ag	10.5	1760
Tin	Sn	7.3	450
Tungsten	W	19.3	6098
Vanadium	V	5.5	3130
Zinc	Zn	6.9	787

Table 4. Fusion Points of Bricks Made from Native Clays[1]

Material	°F
Alumina, fused	3182–3722
Chrome	3542–3992
Clays high in alumina	3276–3416
Magnesite	3992
Silica	3090
Mullite	3236–3272

[1] F. H. Norton, "Refractories," 2d ed., p. 403, McGraw-Hill, 1942.

HEAT-TREATMENT OF HIGH-SPEED STEELS[1]

TABLE 5. TUNGSTEN HIGH-SPEED STEELS

Approximate Composition

Steel No.	C	W	Cr	V	Mo
1	0.70	18	4	1	
2	0.80	18	4	2	0.70
3	0.75	14	4	2	0.50

Heat-treatment

Steel No.	Preheat temperature, °F.	Hardening temperature, °F.	Quenching medium	Tempering temperature, °F.
1	1450–1600	2250–2350	Oil, air, or molten bath	1000–1150
2	1450–1600	2300–2350	Oil, air, or molten bath	1000–1150
3	1450–1600	2200–2300	Oil, air, or molten bath	1000–1150

TABLE 6. MOLYBDENUM HIGH-SPEED STEELS

Approximate Composition

Steel No.	C	W	Cr	V	Mo	Co
1	0.80	1.50	4	1.00	9	
2	0.80	....	4	2.00	9	
3	0.80	1.50	4	1.50	9	5

Heat-treatment

Steel No.	Preheat temperature, °F.	Hardening temperature, °F.	Quenching medium	Tempering temperature, °F.
1	1400–1500	2150–2250	Oil, air, or molten bath	950–1100
2	1400–1500	2150–2250	Oil, air, or molten bath	950–1100
3	1400–1500	2200–2250	Oil, air, or molten bath	950–1100

[1] "Metals Handbook," American Society for Metals, Cleveland, pp. 1000–1003, 1948 edition.

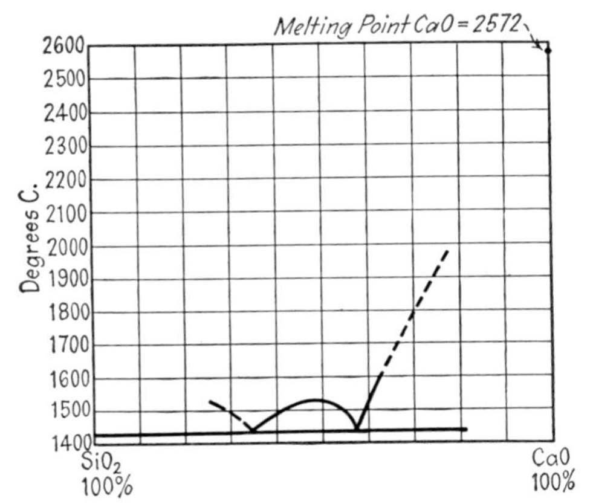

Fig. 126. Approximate liquidus line of part of the SiO_2-CaO (silica-lime) system. Inserted to show that mixtures of oxides have lowering of melting point and eutectic characteristics as well as mixtures of metals. It illustrates the reaction between a lime slag and a siliceous furnace lining. The lining is attacked because the addition of either one to the other lowers its melting point.

DEMONSTRATION OF CRITICAL POINTS

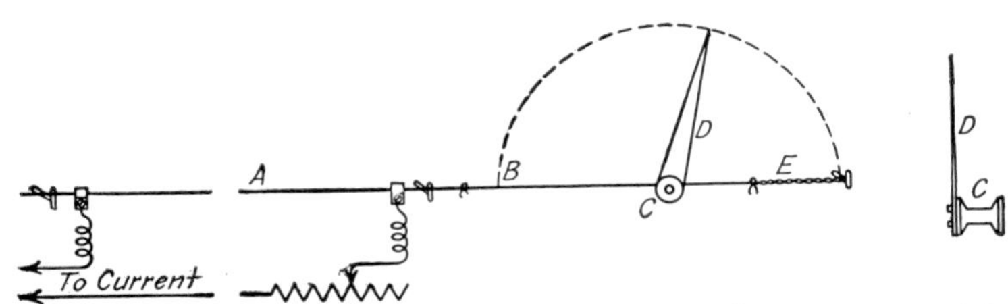

Fig. 127. Sketch of setup for demonstrating critical temperature points.

A very simple demonstration of what happens at a critical point may be observed from the apparatus sketched above.

A is a piece of piano wire, 0.03 in. in diameter and 4 to 6 ft. long, with carbon content near the eutectoid ratio. It is held in tension by a rubber band *E*, a cord *B* connecting them. This cord makes a turn around a common thread spool *C*, to which a pasteboard pointer is tacked. Thus any movement of the cord is magnified by the pointer. There should be a sheet of drawing paper for marking points on the described arc.

Leads from a 110-volt circuit are connected to each end of the piano wire with a suitable rheostat in circuit. The apparatus should be set up in a room that can be easily darkened so that the color of the wire when heated may be observed.

Turn on a current, which should be adjusted to heat the wire slowly. As the wire heats, it expands. The rubber band takes up the slack, the spool revolves, and the pointer begins to describe a circle. It moves uniformly until the lower critical point is reached. At this point it hesitates and moves backward while the temperature is going through the critical range and then resumes its forward motion. At the same time that the pointer was backing up, the wire darkened as if a shadow had passed over it.[1] This period marks the transformation of pearlite into austenite. The darkening was caused by the fact that energy was required for the transformation, and this energy was subtracted from the current that had been heating the wire. As soon as the wire has resumed its regular forward course, it is at the right temperature for quenching for hardening. The current should be shut off very soon to prevent softening of the wire.

As the wire is cooling, an even more marked change takes place. It is startling to see the wire suddenly brighten[2] (at the same time the pointer is *arrested*, made a forward swing, and then resumed its backward motion). The microstructure of the wire has changed from austenite to some structure such as pearlite, sorbite, troostite, or martensite—according to its rate of cooling. The experiment may be repeated several times, and

[1] This phenomenon is known as *decalescence.*

[2] Known as *recalescence.*

finally, during the last cycle, the effect of quenching may be demonstrated.

Just as the current is switched off (be sure that it is) and *before* the recalescence, one end of the wire may be chilled by grasping with a wet rag. Immediately *after* recalescence the other end may be chilled. The latter will be soft, and the former will break at the least attempt to bend it.

SAE NUMBERING SYSTEM[1]

Compositions that do not conform to the SAE compositions, or that are not included in the SAE Standard, should not bear the prefix "SAE."

> A numerical index system is used to identify the compositions of the SAE Steels, which makes it possible to use numerals on shop drawings and blueprints that are partially descriptive of the composition of material covered by such numbers. In the original conception, first digit indicated the type to which the steel belongs; thus '1—' indicates a carbon steel; '2–' a nickel steel and '3–' a nickel chromium steel. In the case of the simple alloy steels the second digit generally indicates the approximate percentage of the predominate alloying element. Usually the last two or three digits indicate the approximate average carbon content in 'points', or hundredths of 1 per cent. Thus '2340' indicates a nickel steel of approximately 3 per cent nickel (3.25 to 3.75) and 0.40 per cent carbon (0.38 to 0.43).
>
> In some instances, in order to avoid confusion, it has been found necessary to depart from this system of identifying the approximate alloy composition of a steel by varying the second and third digits of the number. Instances of such departure are the steel numbers selected for several

[1] Reprinted with permission from the 1949 SAE Handbook.

of the corrosion and heat-resisting alloys, and the triple-alloy steels.

The basic numerals for the various types of SAE steel are

Type of Steel	Numerals (and Digits)
Carbon steels	1xxx
Plain carbon	10xx
Free cutting (screw stock)	11xx
Manganese steels	13xx
Nickel steels	2xxx
3.50 per cent nickel	23xx
5.00 per cent nickel	25xx
Nickel chromium steels	3xxx
1.25 per cent nickel, 0.60 per cent chromium	31xx
1.75 per cent nickel, 1.00 per cent chromium	32xx
3.50 per cent nickel, 1.50 per cent chromium	33xx
Corrosion and heat-resisting steels	30xxx
Molybdenum steels	4xxx
Carbon molybdenum	40xx
Chromium molybdenum	41xx
Chromium nickel molybdenum	43xx
Nickel molybdenum; 1.75 per cent nickel	46xx
Nickel molybdenum; 3.50 per cent nickel	48xx
Chromium steels	5xxx
Low chromium	51xx
Low chromium (bearing)	501xx
Medium chromium (bearing)	511xx
High chromium (bearing)	521xx
Corrosion- and heat-resisting	51xxx
Chromium-vanadium steels	6xxx
1 per cent chromium	61xx
Triple-alloy steels	
Ni 0.40–0.70, Cr 0.40–0.60, Mo 0.15–0.25	86xx
Ni 0.40–0.70, Cr 0.40–0.60, Mo 0.20–0.30	87xx
Ni 3.0–3.50, Cr 1.0–1.40, Mo 0.08–0.15	93xx
Ni 0.30–0.60, Cr 0.30–0.50, Mo 0.08–0.15	94xx
Ni 0.40–0.70, Cr 0.10–0.25, Mo 0.15–0.25	97xx
Ni 0.85–1.15, Cr 0.70–0.90, Mo 0.20–0.30	98xx
Silicon-manganese steels	9xxx
2 per cent silicon	92xx
Low alloy, high tensile	9xx

Type of Steel	Numerals (and Digits)
Austenitic steels (Ni-Cr)	303xx
Castings Steel,	
Corrosion-resistant	60xxx
Heat-resistant	70xxx
Carbon and low alloy	00x, 00xx
High strength	01xx

The AISI system of numbering is basically the same.

THE JOMINY END-QUENCH HARDENABILITY TEST

This test for hardenability has come into such general use that it at least deserves a brief description in this text. It is based on the original work by W. E. Jominy and A. L. Bogehold as reported in the Transactions of the American Society for Metals, vol. 26.

The test is made on a bar 1 in. in diameter and 4 in. long, suspended in such a manner that its lower end can be quenched by a stream of water so regulated that the water does not reach the sides but thoroughly flushes the end.

It is specified that the bar shall be forged or rolled and given a specified normalizing treatment before the final machining to 1 in. in diameter. It is then heated to the proper quenching temperature and quenched as described above. Not more than 5 sec. should elapse between taking the bar from the heating furnace and applying the water, and it must remain in the quenching fixture for at least 10 min.

After removal from the fixture, flat surfaces, not less than 0.015 in. deep, shall be ground lengthwise on opposite sides of the bar, and along these surfaces hardness readings by Rockwell shall be taken every $\frac{1}{16}$ in.

A curve is then drawn, with $\frac{1}{16}$-in intervals as abscissas and hardness values as ordinates. The values on such a curve have been correlated with a great deal of useful information, such as the hardness of various diameter bars at any point between the center and surface; the hardness of several shapes differing from the round; the speed of quench, and much other data valuable in selecting steels and determining heat-treatment.

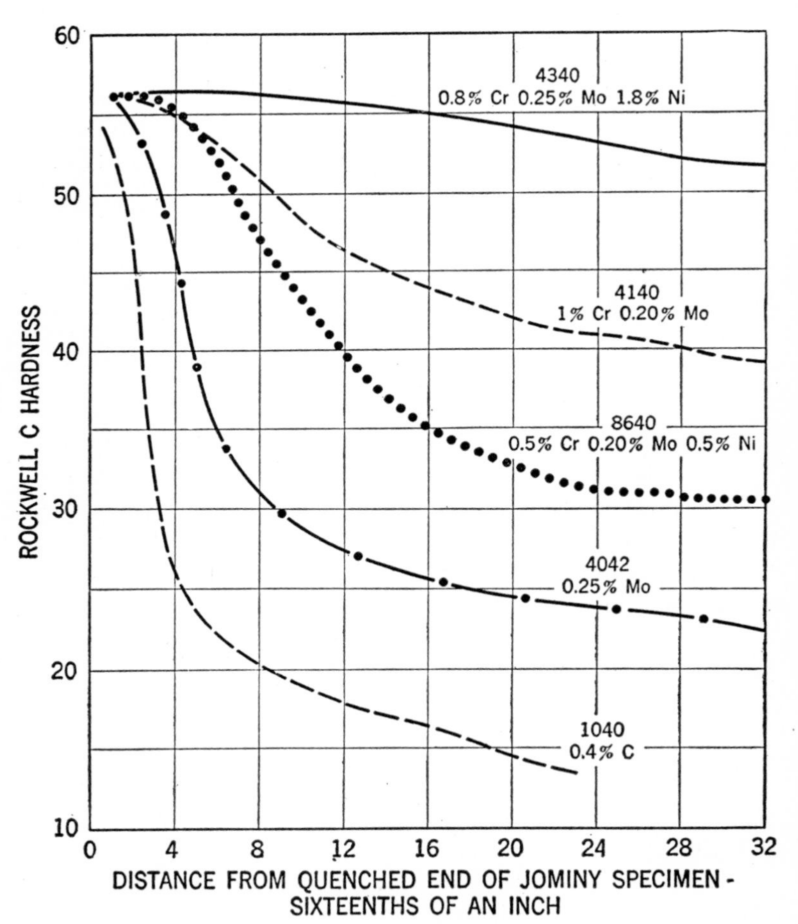

FIG. 128. Jominy end-quench hardenability curves for various 0.40 per cent carbon steels. (*Courtesy Climax Molybdenum Company.*)

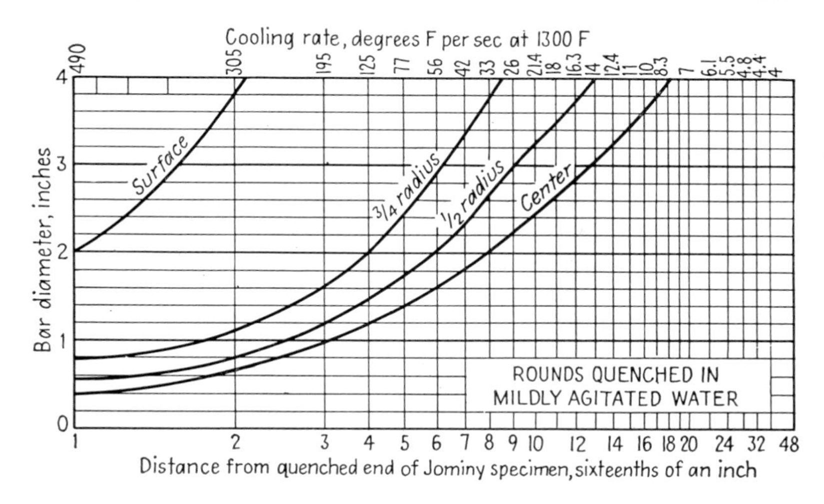

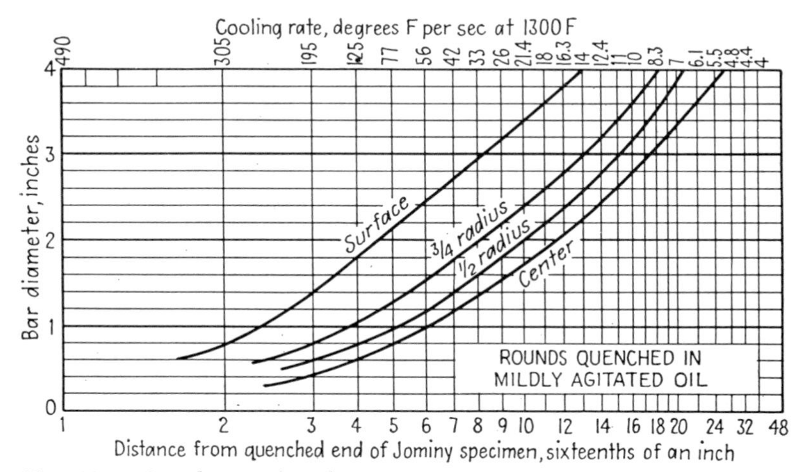

FIG. 129. Correlation of cooling rates in Jominy specimen and quenched round bars (7). (*Reprinted by permission from Society of Automotive Engineers' Handbook, p.* 376, 1949.)

Color Code for Marking Steel Bars

Simplified Practice as Recommended by United States Department of Commerce

Plain-carbon Steels

1010–15, X1015	White
1020, X1020	Brown
1025, X1025	Red
1030–35	Blue
1040, X1040	Green
1045, X1045	Orange
1050	Bronze
1095	Aluminum

Free-cutting Steels

1112, X1112	Yellow
1120	Yellow and brown
X1314	Yellow and blue
X1315	Yellow and red
X1335	Yellow and black
X1340	Yellow and black

Manganese Steels

T1330	Orange and green
T1335	Orange and green
T1340	Orange and green
T1345	Orange and red
T1350	Orange and red

Nickel Steels

2015	Red and brown
2115	Red and bronze
2315–20	Red and blue
2330–35	Red and white
2340–45	Red and green
2350	Red and aluminum
2515	Red and black

Nickel-chromium Steels

3115–20	Blue and black
3125	Pink
3130–35	Blue and green
3140, X3140	Blue and white
3145	Blue and white
3150	Blue and brown
3215–20	Blue and purple
3230	Blue and purple
3240–45	Blue and aluminum
3250	Blue and bronze
3312–25	Orange and black
3335–40	Blue and orange
3415	Blue and pink
3435	Orange and aluminum
3450	Black and bronze

Molybdenum Steels

4130	Green and white
X4130	Green and bronze
4135	Green and yellow
4140–50	Green and brown
4340–5	Green and aluminum
4615–20	Green and black
4640	Green and pink
4815–20	Green and purple

Chromium Steels

5120	Black
5140–50	Black and white
52100	Black and brown

Chromium-vanadium Steels

6115–20	White and brown
6125	White and aluminum
6130–35	White and yellow
6140	White and bronze
6145–50	White and orange
6195	White and purple

Tungsten Steels

71360	Brown and orange
71660	Brown and bronze
7260	Brown and aluminum

Silicon-manganese Steels

9255	Bronze and aluminum
9260	Bronze and aluminum

MOTION PICTURES AND FILMSTRIPS

The visual materials listed below and on the following pages can be used to supplement the material in this book. We recommend, however, that each film be reviewed before using in order to determine its suitability for a particular group.

Both motion pictures and filmstrips are included in this list of visual materials, and the character of each one is indicated by the self-explanatory abbreviations "MP" and "FS." Immediately preceding this identification is the name of the producer; and if the distributor is different from the producer, the name of the distributor follows the name of the producer. Abbreviations are used for names of producers and distributors, and these abbreviations are identified in the list of producers and distributors (with their addresses) at the end of the bibliography. In most instances, the films listed in this bibliography can be borrowed or rented from local or state 16mm film libraries. Unless otherwise indicated, the motion pictures are 16mm sound films and the filmstrips are 35mm silent.

Film users who wish to have information on new films should periodically examine *Educational Film Guide*, a yearly catalog with quarterly supplements published by the H. W. Wilson Co., New York, N.Y. The *Guide*, a standard reference book, is available in most college and public libraries.

Alloy Steel: A Picture of Controlled Development (Bethlehem; MP, 43 min) Present-day steelmaking practices, processing, heat treatment, mill injection, laboratory experimentation.

Alloy Steels: A Story of Their Development (USBM; MP, 20

min) Development and production of alloy steels from 1742 to present day. (Produced in cooperation with the Climax Molybdenum Co.)

Charging and Operating a Cupola (USOE/UWF; MP, 14 min; FS, 44 fr) How to recognize the end of a heat; procedures for dropping bottom and for preparing a cupola for its next heat.

Drama of Steel (USBM; MP, 30 min) History of steelmaking from charcoal furnaces of the ancients to today's blast and open-hearth furnaces.

Golden Horizons (Ampco; MP, 33 min) History and development of copper-base alloys and aluminum bronze.

Hardness Testing: Rockwell (USOE/UWF; MP, 18 min; FS, 49 fr) Need for hardness testing; how to set up the Rockwell Hardness Tester; select and seat the penetrator; select and mount the anvil; test the accuracy and adjust the timing of the machine; and test flat and curved surfaces.

Heat Treatment of Aluminum: Part 1 (USOE/UWF; MP, 19 min; FS, 47 fr) Purpose of heat treatment; microstructure changes; aging or precipitation hardening; effects of heat treatment on the physical properties of aluminum. (Engineering series)

Heat Treatment of Auminum: Part 2 (USOE/UWF; MP, 24 min; FS, 41 fr) Nature of cold, working operations; microstructure changes during cold-working and during annealing; cold-working and annealing operations. (Engineering series)

Heat Treatment of Steel: Elements of Hardening (USOE/UWF; MP, 15 min; FS, 40 fr) How steel is quench-hardened; how the structure and hardness of steels with different carbon content change at progressive quench-hardening stages; an iron-carbon diagram. (Engineering series)

Heat Treatment of Steel: Elements of Surface Hardening

(USOE/UWF; MP, 14 min; FS, 36 fr) Shows how steel is pack and gas carburized; a thin, hard case obtained by cyaniding; nitriding used to obtain a very hard case; how steel is flame- and induction-hardened. (Engineering series)

Heat Treatment of Steel: Elements of Tempering, Normalizing, and Annealing (USOE/UWF; MP, 22 min; FS, 31 fr) How steel is tempered; how the structure, toughness, and hardness of plain carbon steel change at progressive tempering stages; how steel is normalized and annealed. (Engineering series)

Highlights in Steelmaking. Part 1: How Steel Is Made (Bethlehem; MP, 42 min) Production of steel from raw materials throughout all operations. Animated drawings illustrate blast-furnace smelting and three steelmaking processes—open-hearth, bessemer, and electric-furnace.

Highlights in Steelmaking. Part 2: Steel Treating and Testing (Bethlehem; MP, 45 min) Processing of steel in mills; heat-treatment of steel—quenching, tempering, normalizing, and annealing; testing of steel.

Making and Shaping of Steel: The Making of Steel (USBM; MP, 15 min silent) Operation of open-hearth furnace, electric furnace, and bessemer process.

Metal Crystals (Am Soc Metals; MP, 33 min silent) Crystalline and noncrystalline substances; microscopic technique with metal specimens; temperature of solidification. (Prepared by Committee on Visual Education of ASM in cooperation with Ohio State University)

Metal Magic (GE; MP, 10 min) Crystalline structure of metals and the development of new alloys.

Molding with a Loose Pattern (USOE/UWF; MP, 21 min; FS, 37 fr) How molding sand is prepared; how to face a pattern; ram and vent a mold; roll a drag; cut a sprue, runner, gates, and risers; swab, rap, and draw a pattern; and what takes place inside a mold during pouring. Animation. (Foundry series)

Powder Metallurgy. Part 1: Principles and Uses (USOE/UWF; MP, 19 min) Principles of powder metallurgy—powder, pressure, heat; major industrial applications of powder metallurgy; laboratory process of combining silver and nickel powders. Animation. (Engineering series)

Powder Metallurgy. Part 2: Manufacture of Porous Bronze Bearings (USOE/UWF; MP, 15 min) Manufacturing process by which metal powders are fabricated into porous bronze bearings and impregnated with oil. (Engineering series)

Preparing a Cupola for Charging (USOE/UWF; MP, 21 min; FS, 48 fr) How to recognize the end of a heat; procedures for dropping bottom and for preparing cupola for its next heat.

Stainless Steel (USBM; MP, 29 min) Qualities of stainless steel; making of stainless steel. (Produced in cooperation with Allegheny Ludlum Steel Corporation)

Tension Testing (USOE/UWF; MP, 21 min; FS, 45 fr) How a hydraulic tension testing machine operates; how to prepare the machine and a specimen for a test; conduct the test to determine the specimen's elastic limit, yield point, and ultimate strength. (Engineering series)

This Moving World (MFS/Assn; MP, 30 min) Preparation of molds, melting of metal, annealing process for converting iron into malleable iron.

X-Ray Inspection (USOE-UWF; MP, 21 min; FS, 54 fr) Use of radiographs in industry; generation of X rays in the X-ray tube; wave nature of X rays; procedure in making radiographs; and interpretation of radiographs for defects in metals. (Engineering series)

SOURCES OF FILMS LISTED

Am Soc Metals

American Society for Metals, 7301 Euclid Ave., Cleveland 3, Ohio

Ampco

Ampco Metal Inc., 1745 So. 38th St., Milwaukee 46, Wisc.

Assn

Association Films, Inc., 35 West 45th St., New York 19, N.Y.

Bethlehem

Bethlehem Steel Co., Bethlehem, Penna.

* Also available from 35 District Offices. Write for catalog—listing addresses of these libraries.

GE

General Electric Co., 1 River Road, Schenectady, N.Y.

USBM

U.S. Bureau of Mines, 4800 Forbes St., Pittsburgh 13, Penna.

USOE

U.S. Office of Education, Washington, D.C.

* Films sold under U.S. Government contract by United World Films. May also be borrowed or rented from those 16mm film libraries which have purchased prints. Write to the U.S. Office of Education for a list of such libraries.

UWF

United World Films, Inc., 1445 Park Ave., New York 20, N.Y.

INDEX

S